COLLABORATION IN THEATRE

MW00810151

ME

COLLABORATION IN THEATRE

A PRACTICAL GUIDE FOR DESIGNERS AND DIRECTORS

Rob Roznowski and Kirk Domer

First published in 2009 by
PALGRAVE MACMILLAN®
in the United States—a division of St. Martin's Press LLC,
175 Fifth Avenue, New York, NY 10010.

Where this book is distributed in the UK, Europe and the rest of the world,
this is by Palgrave Macmillan, a division of Macmillan Publishers Limited,
registered in England, company number 785998, of Houndmills,
Basingstoke, Hampshire RG21 6XS.

Palgrave Macmillan is the global academic imprint of the above companies
and has companies and representatives throughout the world.

Palgrave® and Macmillan® are registered trademarks in the United States,
the United Kingdom, Europe and other countries.

ISBN-13: 978–0–230–61421–5 (paperback)
ISBN-10: 0–230–61421–3 (paperback)
ISBN-13: 978–0–230–61420–8 (hardcover)
ISBN-10: 0–230–61420–5 (hardcover)

Library of Congress Cataloging-in-Publication Data

Roznowski, Rob, 1963–
    Collaboration in theatre : a practical guide for designers and
    directors / Rob Roznowski and Kirk Domer.
        p. cm.
    Includes bibliographical references.
    ISBN 0–230–61420–5 (alk. paper)
    ISBN 0–230–61421–3 (alk. paper)
    1. Theaters—Stage–setting and scenery.
  2. Theater—Production and direction. I. Domer, Kirk. II. Title.

PN2091.S8R69 2009
792.02'32—dc22   2008034625

A catalogue record of the book is available from the British Library.

Design by Newgen Imaging Systems (P) Ltd., Chennai, India.

First edition: March 2009

D 10 9 8 7 6 5 4

Printed in the United States of America.

This book is dedicated to Jillian Blakkan-Strauss, our undergraduate researcher who helped us invaluably.

CONTENTS

Part II   Collaboration in Practice

Part III   Collaboration in the Classroom

FIGURES

Abstract

Objective: Define a successful collaboration between directors and designers.

Premise: Defining and nurturing communication within a production team leads to a successful collaborative process.
This book is divided into three parts:

Part I: Collaboration in theory—part I will present directors and designers with guidelines for creating a successful theatrical collaboration.

Part II: Collaboration in practice—part II will trace the successes and failures of a fully realized production that used the theories presented in part I.

Part III: Collaboration in the classroom—part III will create a course drawn from theory and practice that teaches the collaborative process.

ACKNOWLEDGMENTS

We wish to thank all students from the Collaborative Studio classes, Dr. Dixie Durr for discussing this book well into her retirement, Dr. George Peters for selfless support, Kate Bushmann for her attention to detail, Martha Bates for publishing guidance, Barry Delaney for testing our theories and all of our theatrical collaborators in the past, present, and future.

INTRODUCTION

COLLABORATION ON A COCKTAIL NAPKIN

One night we were out at a bar unwinding from school and we began talking about an upcoming show . . .

K: My memory of how we got the title for this introduction is in the context of a show you were directing but I wasn't designing, Evan Smith's *The Uneasy Chair*. I hadn't read the play, but you needed a sounding board. So you said things like, "Well, how would you do *this*? Because I need *this* to happen *here*, and I want the feeling of *this* in that scene. And I have to have *this* so it can split and move apart and come back together . . . ." And I was confused. So you started drawing it on a cocktail napkin. And as you drew I was saying, "Well, what if you added this? Watch your symmetry. Is that really what you're trying to say about the show?" Then the napkins unfolded as we expanded. Most of what we worked on never came to light, but the essence of our give and take was constructive. And that is exactly what you do as a director and designer.

R: I had some preliminary ideas, and you transformed them beyond what my vision was. Someone who had no connection to the play kept asking the right questions and pushing my interpretation into a clear design . . . I found that moment really special.

K: That moment was *great*!

R: There was no judgment in this. From either of us. We were just bouncing ideas off of each other as peers, as collaborators, without worrying whether the design was going to get produced. It was just questions and answers and refining.

K: It helped us create something that we didn't end up using. But it allowed us to see the way each other worked.

R: It was fun to witness your creative process. It was exciting to watch an artist maneuver my analysis and turn it into a picture. A world, really. I can't do that.

K: That's why you need me.

R: What happened on those cocktail napkins captures those rare moments when designer and director think as one. When you finish drawing and sharing ideas, it's hard to remember how you got there. At the end you can look at all the napkins and try to claim your individual ideas throughout, but that final napkin is ours!

K: I think that "collaboration on a cocktail napkin" implies that creative moments can come at any time. You have to be willing to collaborate anytime and anyplace.

PART I

COLLABORATION IN THEORY

The word "collaboration," used by theatrical production teams* made up of directors and designers, has been under much debate recently. Some think collaboration has a negative connotation, as conspiring with the enemy. Others prefer the word "cooperation."

To us, cooperation in a production team implies a handshaking group of individuals promising not to tread on one another's toes. Our favorite analogy is a co-op store, which doesn't sell stew; it sells carrots, potatoes, and meat—all separate ingredients. We are not interested in working that way. We want the stew.

Collaboration implies a meshing of ideas to us. A production team is comprised of separate individuals who indeed cooperate with one another, but also inspire and affect each other to produce a cohesive production. We hope to improve how directors and designers arrive at a mutually satisfying creative experience by illustrating through theory, practice, and example a better way to collaborate.

"Collaboration" or "Cooperation." It all comes down to vocabulary.

Note

*For our purposes, the term "production team" refers to the director working with the set, costume, and lighting designers. Of course, other important contributors—such as the sound designer, technical director, choreographer, stage manager, and many more—may be part of your team. The ideas in this book hold true for any member of your team.

CHAPTER 1

THE VOCABULARY OF COLLABORATION

This chapter offers ways to assist in dealing with the initial meetings of the production team. The first impressions that you present to your collaborators are difficult to alter. Although we are using examples from our lives working at a university and in professional settings, our experiences can easily translate into any production venue. Our aim throughout is to assist in creating a healthy collaboration. Also, the practical part of this book aims to model an expeditious collaboration and therefore places the director as arbiter of most debates. This should not always be the case.

This chapter, regarding first meetings of the production team, will

- stress the importance of shared vocabulary for designer and director;
- offer words to use and to avoid;
- define a well-chosen vocabulary equal in its respect to designer and director;
- teach ways to avoid future misunderstandings by adopting a common vernacular;
- define ways in which a director and designer work;
- clearly delineate a chain of command;
- offer alternate methods of working when someone does not speak your "language";
- help the collaborator explore past experiences to find more successful ways to collaborate;
- pave the way for a successful collaboration between designer and director; and
- define Collaboration.

Shared Vocabulary

A production team comes together in the professional world, the realm of academia, in a community theatre, or in the high school auditorium. In all instances each team member brings a unique level of education and experience, and therefore a production vocabulary culled from past collaborations. The objective is usually the same: to put on the best show ever.

Collaborating with theatre artists from different backgrounds should be a beautiful way to educate oneself, to adapt one's process to fit another's, or to expand one's technique. In practical application, however, distinct differences within a design team can lead to frustration, miscommunication, disrespect, and a theatre full of bruised egos.

The collaborative process—it sounds so peaceful, so unified, so simple. Then, why is it some production meetings end in tears or raised voices?

VOCABULARY.

In most cases, those high emotions are due to verbal miscommunication. Very few intend to hurt a fellow theatre artist unless they feel their ego has been attacked.

R: A design colleague chastised me when I called an interior box set a "simplistic" design in comparison to larger-scale opera and musical designs. This designer told me how "naïve" I was and that I was "just like every other director in thinking that all of those details are simple." Did I mean to be antagonistic or diminish his work? By all means, no! Did I mean to imply that a one-setting interior is easier than the twenty-three-scene musical we had just worked on? Yes! After explaining my side of the matter, things calmed down. That person is the co-author of this book.

K: I remember the story as such: We were selecting our production season. We were figuring out the order in which the plays would be presented. Someone (my co-author) said, "Put that play in this spot. It's just a box set. That'll be easy." That person implied that those realistic minute details are easy. It is *complex* in nature. Just as in operas, just as in musicals, just as in any production, each play has its own complications.

So, no matter how well you think you are communicating, how good your intentions are, or how respectful you are to one another, sometimes words just get in the way.

Director Says:	Designer Says:
I need levels!	I can't afford it!
You aren't getting what I am saying.	Then I need clarification.
What if we just put this there?	There's not enough room!
Can we just make it brighter?	I am out of instruments!
She looks fat in that.	That's not my fault!
Well, in the original production...	Are we doing the original production?
I can't work like this!	I'm on my break!

Although most theatre artists have used some semblance of the above statements, following a few simple steps may ease the collaborative process.

Getting the Job

As the production team gathers, the collaboration begins. The initial meetings are difficult. Your insecurities emerge. You don't feel qualified. You don't know if you can do this production, no matter what your track record or successes. It is the same for director or designer. The producer calls for an interview, and a million reasons why you aren't the right person for this show creep into your mind. To avoid such creeping, follow this *rule*:

DO A BACKGROUND CHECK.

That means research the company, school, or person hiring you, in addition to researching the show about which you have been contacted. Researching the producing organization or person contacting you will allow you to give a focused and relaxed interview. An organization's mission statement is always a good place to start. You have created the base for a healthy collaboration. In researching the producing organization you may find a unique hook that corresponds beautifully with this team. In all cases, background work makes for a fruitful, more challenging, and a truer discussion than the normally awkward first interview.

Providing you have done your background work, start making correlations. You don't create in a vacuum; you have experience, knowledge, and taste to help you hone your discussion. Find the common ground in your previous work that makes you right for the show rather than wrong. Find successes and failures in productions that may be excellent talking points. If you have never directed or designed a Shakespeare play, why not say, "I really want to expand my resume/portfolio?"

> *R*: OK, I once backed out of directing a production by lying about a family member's health right before rehearsals began. I just couldn't see that the production would be successful. I bailed because I knew it would fail...I wrote my own reviews. If the producers are reading, I apologize.
> *K*: I have the opposite problem. Rather than talking myself out of a production, I feel like I have to take EVERY design. I tell myself I have to expand my connections and am always thinking I might miss out on something. Until very recently, I did not have the vocabulary necessary to say NO.

Adapting Your Style

The interview went well. Your homework paid off. You enter the collaboration.

As designer or director, be sure to listen for the repeated phrases used by the producer or company manager. Once you hear these phrases used often, you should begin to experiment with them yourself. If "world of the play" keeps coming up, you had better say, "In helping define the world of the play, I would like to talk about the world outside of the living room." Or if the producer keeps speaking of the educational aspect of the show, mention your master class with high school students. This brings us to the next *rule*:

ADAPT YOUR VOCABULARY.

By focusing your attention on the vocabulary of your collaborator, you may begin to create a common, expanded language. The clarity of language must not be taken for granted. Everything needs to be spelled out.

How does a director do that while not appearing to be a condescending micromanager who distrusts herself or her designers? How does a designer give enough information without inundating the director with superfluous details? In initial meetings, set up clear expectations through example. By proving to be a well-spoken member of the production team, you are making yourself an indispensable part of the interpretation. Clearly expressing your position secures your role in the creative process. A director who runs a thorough first meeting will model how detailed everyone's work should be. A designer who can speak intelligently about script analysis becomes a prime collaborator.

In many theatre training programs or geographical regions, there are specific terms that have been created or adopted through necessity, private jokes, or unique history: a theatre-specific shorthand. If you don't know a term, don't feel awkward about asking what that means.

> R: Each time I visit acting students at another school, I adapt to the acting terminology used in that program. Do I want to spend the time trying to force the students to refer to "objectives" instead of "goals" or "actions?" No, let's get the work done. We are all working toward the same goal.
>
> K: When I work in different parts of the country the same thing is true, but in some cases I even have to adjust to specific climate conditions. Due to the humidity, it is better to paint a drop vertical in one part of the country and on the floor in another. Whatever way it is painted, we combine efforts, learn to communicate, and ultimately get the show done!

If you do find new terms that you like, adopt them. When entering into a new theatre, why not explore their vocabulary? If it seems foreign, try to engage in a conversation about vocabulary. We don't mean to imply that you alone must adapt your language. Do not hesitate to introduce others to your own vernacular. Remember, if you are married to your specific terminology, please be sure to define it for others.

Clarifying Definitions

Once you have had the opportunity to hear the vocabulary of your director or designer, you may begin to adopt these words only if you fully understand them. For example, "You'll be getting a model from me." Does that mean a white model, color model, or virtual model? When someone says, "You'll have sketches at the next meeting," make them define it. Will these sketches be watercolor or thumbnails? The next *rule* of collaboration:

ALWAYS ASK FOR THE DEFINITION.

Even if the word being bandied about is something you think you understand, ask your collaborator to define it if there is the slightest confusion. You can avoid so many misunderstandings by getting to the collective meaning.

If neither the designer nor director is speaking your language, you may find yourself setting the vocabulary standard. A designer has as much right as a director to ask someone to be more specific. If the director says Shakespeare's *Romeo and Juliet* is about love, you are perfectly within your rights to ask him to define this generalization. Let's say the director responds with, "All kinds of love—tragic, romantic, young love!" If this is too general (and it is), ask for more explanation. "What specifically about tragic, romantic, young love?"

R: We were involved in a master of fine arts [MFA] student's committee meeting where a designer complained that her director had presented no clear concept. We charged her in the future to needle the director until she got a clarified premise she could hook into. "But this director doesn't like to talk like that," she replied.

K: If you are working with someone who really cannot speak your language, try changing up the idiom. Bring in music, food, or poetry to ensure that you can relate your work to something the director points to. The director may hook into something you brought.

By understanding the way in which *you* as designer work, you may influence a director to work collectively at the next meeting. Bring in a few premise statements (explained in part I, chapter 2) written by you, bring in evocative images, or found objects to force the director to give you more guidance. Beg the director to at least define the style. Simply get down to the basics. Is this farce or satire?

Let's say in that preliminary meeting the director is not amenable to the above and hated all of your imagery, food, and music. If a director will not give you the necessary guidance, it is up to you to inundate the individual with a large number of rough sketches. Don't spend much time on the mechanics. Bring in lots of options and let the director live with them. When you next meet, the director should have narrowed the choices, and you can begin to refine. Initial ideas are exactly that: initial. The next *rule* of the initial meeting:

ONLY PROCEED WHEN IN AGREEMENT.

Avoiding Buzzwords

Each production entity has its own shorthand and/or buzzwords. It's sometimes frustrating being left out of the loop when everyone speaks his own special vocabulary—even if it is in clichés. So, instead of rolling your eyes when the buzzwords start flying (e.g., "Think outside the box"), force them to define it. Where is "outside?" And what exactly is "the box?"

The reverse is true for you as well. All of us are certainly guilty of repeating phrases to save time in discussions. So, to ensure a smooth collaboration, do some mock interviews or presentations. Record them and then watch and critique yourself. What are the phrases that you keep using? On that recording you may notice some annoying habits that may get in the way of securing a second meeting. Open-ended words like "environment" should be scrapped when meeting someone for the first time. By avoiding generalizations, you will be forced to find new definitions for each production and for yourself. Understanding your contribution to the collaboration can make it a deeper experience for all.

RULE: AVOID BUZZWORDS.

Including using the word "buzzwords." Then you will truly be outside of the box.

Respecting the Chain of Command

Vocabulary related to the chain of command is a difficult and territorially sensitive challenge. By following the previous rules and doing your research, you may answer the questions that plague most collaborations. From the first moment of the first meeting you must define your role in the collaborative process. You are in complete control of your role in this union of theatre artists.

RULE: ESTABLISH A CLEAR CHAIN OF COMMAND.

If you do not offer your opinions, you may not complain that no one listens to you. If you are so vocal that you ignore everyone else's opinion, you can't cry about doing all of the work alone. A collaborator must somehow enter into the shared vocabulary while respecting the chain of command.

To create the practical collaboration in the title of this book, designers must unite their creative choices with the vision of the director. And in turn, each must funnel his or her work through the text of the playwright. After much (healthy) debate between the writers of this book, we agree that the usual chain of command for a production team begins with the script and flows directly into that of the director (for interpretation) and finally to each of the designer for their individual specialty (i.e., script → to director → to designer).

Eventually you are aiming for a more equitable and layered chain of command, where ideas flow freely from one collaborator to the next with the production serving as the hub. Imagine a wheel with the production at the center and the script, director, and designer acting as the spokes. This sort of exchange comes from trust and experience.

The first option implies a committee where the director is the chairperson. The committee discusses an issue and the chair has veto power. The second option implies that all participants are on an equal playing field and the production holds veto power. The first is practical while the second is ultimately more satisfying yet difficult.

> R: Our faculty discussed with an MFA lighting design candidate that he might be too laid back in the design process. He answered that he was a laid back person. He liked coming in after the costumes and sets were completed to do his work. He liked working around the world of what others had created. He chose this role in the design process and he was fine with it.
>
> K: That student had no complaints about his contribution. He formulated his own chain of command. It might be frustrating for others involved, but it worked for him.

In most cases the director is the prime interpreter of the script. As described in this chapter, the director communicates a clear premise to the production team then moderates the ensuing artistic debate. This debate between all involved can enhance, modify or significantly alter the director's original interpretation.

The first job of the director is to create a clear vision for the production by describing his own premise using a vocabulary that respects the contributions of everyone involved. He also models the work ethic and generates the dynamic of the production—from technician to actor to crew to house manager. This approach serves a dual purpose: not only does the director need his cohesive ideas communicated, but he also needs to provide the environment to support the best performance of the designer.

It may seem like we are discarding any sort of original ideas by the designer and forcing them to conform to the director's whim. Absolutely not. A designer responds to the director's vision using his or her own talents. With a confident director, a designer may feel free to push the vision in new directions with unique ideas. If a director without a clear idea is at the helm, the designer may easily guide the production.

Another form of collaboration that is completely valid is when a design team meets and simply says, "What should we do?" The ideas yielded from this conversation have the ability to create truly stunning work. The issues that can occur in these instances are that this process is more time-consuming and allows for more bruised egos if an idea is rejected. This sort of collaboration is specialized and sometimes best left to teams that have a history of trust and respect. For our purposes, we remain focused on the time-efficient model of collaboration mentioned earlier.

> *R:* I witnessed a production where a director had no idea what the play was about. The designer came in with a concept before they had even spoken. The director having not established himself as the top of the chain of command allowed himself to be pushed aside. The production ended up looking exactly like that—a conceptual production that was not cohesive.
>
> *K:* As advisor of this production, it was difficult to watch the process when the student designer was leading the pack because in the world of academia sometimes we learn best from our mistakes.

Implicit in most of this book is the assumption that a director will arrive at meetings with a concept in mind. This is not always the case. Obviously not all productions require heavy "concepts"; they simply require the production. The director should still guide the message of the show, but it is just as valid for designers to come in and offer design and premise elements for the production. At initial meeting a director may simply ask: "So what do you think? What should we do?" Designers should be just as prepared for this conversation. The director, as arbiter of suggestions from everyone, remains the constant, but in this instance all suggestions are welcome. This is collaboration.

Respecting Your Fellow Artists

The talent and ego of an artist (whether designer or director or actor) is really all the artist has. It is this individual's contribution that serves the

production. So by dismissing someone's sudden idea, brushing off t
research, or mocking her vision for the show, there are bound to t
hurt feelings; and rightly so, especially in these early meetings. If so
involved in the production is doing her job, you should respect that. ...y ιο
follow this *rule* throughout the process:

≱ RESPECT AND CONSIDER EACH IDEA.

Of course, this does not mean you have to agree with every idea. And certainly each idea merits different amounts of consideration. BUT, don't dis-
miss anything. That particular idea may blossom into a great discussion,
leading to a revelation concerning the production.

> R: I recall myself as a neophyte director so secure (yet insecure) that I had
> the answer to every element of my first few shows. I was unable to allow
> anyone in on my vision. The designers were facilitators of my ideas. It
> wasn't until I acquired more trust (in myself and others) that I could
> allow a designer to alter or enhance MY vision. It is a great moment
> when a designer comes up with a BETTER solution that serves the show.
> It makes the director feel SO much less lonely.
>
> K: I find myself working with a lot of neophyte directors, whose sole direc-
> tion is to recreate a metaphor or image. Nothing can sway them from
> THEIR lone image no matter how many possibilities I offer. That
> inflexibility forces the designer to become a facilitator.

You may be surprised what can come out of a brainstorming session with
many artists concentrating on a specific challenge in a production. The
respect and consideration begins in the first meeting with designer and
director. So choose your words carefully in these meetings. Once trust has
been broken, the road to recovery is difficult. The disgruntled tone of the
production team obviously creates a similar production.

Communicating Positively

What Directors Like to Hear from Designers

Rather than Saying:	Say:
No.	Let me think about it.
We won't have time.	Can you prioritize?
It's too expensive.	Here are three alternatives.
I didn't get as much done as I wanted to.	Here's my plan with deadlines...
You're just going to have to compromise.	Have you thought about these options?
You don't understand it from my point of view.	How does this all fit into the premise?

For the most part, when a director says, "How's it going?" they really just want to hear "Great!" And while directors want to be involved in large discussions, they want to serve as supervisor rather than manager.

What Designers Like to Hear from Directors

Rather than Saying:	Say:
I need levels.	The movement requires playing areas.
The scenery should be funny.	How about unusual traffic patterns?
Brighter!	I am having trouble seeing her reaction.
I want to keep it simple.	Why not let the text be the focus here.
Just design the show and I will use it.	Let me see what you can come up with.
Move that bench downstage.	Can you help in motivating this cross?

For the most part, when a designer says, "What do you think?" they really want to hear "Great, and…" Designers want to be involved in the overall process as well as develop their own design to its fullest. They need time to work alone based on discussions and presentations.

Collaboration Begins…

- It is the free flowing exchange of ideas without ego or resentment
- It is the moment when ONE idea becomes a group's rather than an individual's
- It is an idea that enhances the production rather than an individual area
- It is the moment when you begin to finish another person's sentence
- It is the moment when you feel justified without justification
- It is the scrapping of superficial politeness for genuine artistic discovery
- It is the moment when you have created something special
- It is the moment of epiphany when the production becomes cohesive
- It is the "Eureka" moment

Summary

As you can see, you can simply rephrase the same point in a respectful and positive manner. In fact, all of the rules in this chapter are simply guiding you to develop a more considerate tone. Only through that mutual respect in these first meetings (and beyond) can artists attain the meshing of styles necessary for the collaborative experience we envision. To review, the *rules* to achieve this include:

- DO A BACKGROUND CHECK.
- ADAPT YOUR VOCABULARY.
- ASK FOR DEFINITIONS.
- ONLY PROCEED WHEN IN AGREEMENT.
- AVOID BUZZWORDS.
- ESTABLISH A CLEAR CHAIN OF COMMAND.
- RESPECT AND CONSIDER EACH IDEA.

By paying careful attention to your language with fellow theatre artists you may establish an arena where all may feel free to collaborate. Of course, these initial meetings are also a chance for any of the collaborators (having exhausted the *rules*) to realize, "I can't work with this person." This is sometimes a very liberating realization. But if you commonly respond this way after your meetings, you may want to take a closer look at yourself.

Oh, one final *rule*: These aren't rules, merely suggestions. The word "rules" often carries a restrictive definition. Each production and each member of the team is unique. Be flexible or assertive when necessary.

CHAPTER 2

SCRIPT ANALYSIS FOR COLLABORATION

F or designers and directors alike, the entire production revolves around one collaborator who rarely attends production meetings: the playwright. You have heard it before but, honestly, it all begins and ends with the script: the common ground from which we all create. The deciding factor in most debates and the duty of any member of the production team is to serve the script. The sole reason for everyone devoting the time and effort to any production revolves around the script—or more correctly, around the production team's *interpretation* of the script.

That word interpretation is where the majority of the collaboration occurs. Collaborating on a collective interpretation is the primary focus of those first design meetings. Arriving at this interpretation is a skill lacking in some collaborators due to many factors that we will explore. This chapter addresses ways to arrive at a "true" collaborative approach to script analysis. We will also offer a few ideas to assist you in creating your own process for script analysis.

This chapter will show you how to

- experience the script in the first reading;
- examine the play for its structure and theme;
- extract a premise;
- prepare for the first meeting; and
- direct and design for the script.

As we wrote in chapter 1 of this part, just as there are several ways to achieve the same goal, there are MANY ways to analyze a script. By taking the time to analyze on your own and then discuss it with your other collaborators, you may achieve a utopian cohesive collaboration. The miscommunication and inflexibility of collaborators regarding script analysis contribute to creating a shaky base of an otherwise solid collaboration. The practical part of this book addresses traditional script analysis but the pointers explored can be used for nontraditional theatrical pieces as well. Analysis lays the groundwork for future decisions on any show.

The Director's Process

Once the play has been selected and your role on the production team is defined, immediately read the script. Immerse yourself in the experience of reading it. Try to keep all design or directorial decisions out of this first reading. Simply read the play and allow yourself to enjoy the piece.

You may have some history with the play. Forget it (at this point). You may have read it before. You may have seen stellar or awful productions. You may have seen the film. You may have heard others' opinions. You may have read some reviews or historical writings. At this point in the process, you must try to push all of that history aside and read the play as an audience member. Not as your subscribers. Not as your peers. Not as your students. Simply read for yourself as the sole audience member.

In this first experiential reading of the play, some initial thoughts usually enter a designer's or a director's mind. You may want to jot them down, but put them away! Those initial thoughts are simply that—initial. Make no firm decisions. The only job you have after your first reading is to examine your gut reaction. Questions may include

- What did I like about the play?
- Where was I confused and why?
- Why does this character resonate with me?
- How does this play relate to me?
- Why is this play relevant?
- What was my favorite/least favorite aspect of the play?
- Who is this play about?
- Why did the playwright pick this title?

As you can see, these are VERY basic questions that examine the audience's (your) reaction. No questions of the play's style, message, or design elements need enter into your first reading. Savor this reading. You owe it to yourself, the script, and your collaborators to truly experience the play.

> R: Imagine when I was asked to direct Richard O'Brien's *The Rocky Horror Show!* Talk about the challenge of avoiding any preconceived notions. It was only when I blocked the movie out of my mind that I truly examined the play's original intent and read the lyrics for their message and not another's interpretation. I had to forget my memories of many midnight

showings and silence Tim Curry's voice in my head. Of course, that history served me well later in the process, but in this initial phase it was all about returning to the text.

K: A couple of years ago I designed Shakespeare's *Twelfth Night* for the THIRD time. What was going to be special about this production? How was I going to divorce any feelings that had resonated with me from one director that were now being overlooked by another? I also had to silence those former directors to collaborate with this new production.

Primary Goals

The primary goal of any director or designer is to draw the audience in and shape their experience as they watch the production. Remember, what moves you about a show is what should move your audience. The introduction of dramatic structure, premise, and other theories within these pages are simply ways to help a collaborator focus or deepen the experience.

Script Analysis

Once you have read the play for the experience, your work begins as you read the play again—this time in the role of critic. "Critic" here means turning a "critical" eye toward exploring the play's theme, not criticizing the play. In this second reading of the script, write down words or phrases that resonate with you. Circle lines from the script that reveal what you think the playwright's message is. (We will return to this idea.)

At the same time, you are also trying to define the play's dramatic structure. Here, we enter a whole new territory of vocabulary issues. The dramatic structure debate begins. We, the authors, have often argued about plays' inciting incidents and climaxes. Although some feel that the highest point of tension is the climax, others feel that a character's turning point makes for the true climax. Are you Aristotelian in your approach, or have you been taught another method? However you analyze a play, you must support your choices with the text.

(If you don't have a keen grasp of dramatic structure, take a look at David Ball's *Backwards and Forwards*.)

Do we think it is imperative that each member of the production team analyzes the play in the same way? No. As we said, we ourselves do not always analyze a play in the same way. But we do insist that the *team* must agree on the dramatic structure so that everyone may "frame" these important moments. If a disagreement with grounded support ensues, the director has the final say. A designer may certainly try to sway the director, but we hope that respectful debate will lead to an amicable final decision.

> R: One of our biggest debates was about David Auburn's *Proof.* Is the climax at the end of Act One when Catherine announces she wrote the amazing proof? Or is it when Catherine finally proves that, though she may be like her father, she will get through life with Hal's help? Two very different climaxes yield two very different productions.
>
> K: This debate came out of my first teaching of an introductory play analysis course. I had always believed that the latter of the two above was the climax, but when teaching three different analysis theories simultaneously, it was exciting to find out that both choices could be supported (the logical climax and the emotional climax).

When entering the collaborative process, you must be flexible and open to another's supported opinions. It is your responsibility to explain your version of dramatic structure through exact moments or lines from the script. Support from this second reading of the text will allow your opinion be heard in the collaboration. It is imperative for all on the team to thoroughly examine the structure.

Creating a Premise

Once you have completed your reading of the script as critic, it's time to collect all of your impressions up to this point and formulate your premise. Consider your dramatic structure with exact lines as support, all of the words you jotted down, and all of the circled lines from the later reading. After reviewing these, begin brainstorming and write down other words or phrases that the play evokes in you. This sort of free association is the time to explore any areas of the play that resonate with you—many of which you may recall from your first reading.

Once you have completed this process, put those ideas and words into a tidy premise for the show. An explanation of "premise" follows, but remaining "tidy" is also important. Tidy here implies that nothing is lost or forgotten—organizing all of those messy subplots and minor characters to create a cohesive and ordered synthesis of the entire show.

The premise is an active statement of what you believe the show is about. This is a VERY watered down version of Lajos Egri's premise ideas in his wonderful book *The Art of Dramatic Writing: Its Basis in the Creative Interpretation of Human Motives*, which is mandatory reading for any theatre person. By creating a premise you will avoid the production of *Romeo and Juliet* that is about "love." The premise helps create a substantive foundation for your production on which every decision can be based.

Mr. Egri goes into much more detail, but the basic idea is that you seek to create a phrase about the show that follows this pattern:

SOMETHING leads to SOMETHING ELSE.

The biggest aid in formulating your statement is defining a main character's journey in conjunction with the climax of the show. These two important clues usually reveal the premise you seek. You do not have to use a premise per se, but you must consider a statement or concept that grounds your production. A production cannot be summed up by simply representing a mood, word, or image—it must have a message.

A premise can change from production to production of the same script. For example, if we decide that a production of Tennessee Williams' *A Streetcar Named Desire* will place Stella as the main character, our premise may be "Balancing husband and family leads to a horrific decision." This production will be about the battle for Stella's support. Stella will be the focus of a fierce competition between Stanley and Blanche for her soul. Ultimately this battle climaxes with Stella's decision to send Blanche away. The design may reflect this by having Stella onstage giving birth to her baby while the rape scene occurs. The climax is a battle for her soul.

If you believe the main character is Stanley, you may say the premise is "Conquering all doubts leads to ultimate satisfaction." You may have a production where Stanley is the focus of EVERY moment onstage, and the main purpose of the show is to highlight the animal brute quality of Stanley. The costumer will place Stanley in the most savage of clothes and the prop designer will make the package of meat that Stanley throws at the top of the show leak blood from the fresh kill. The climax between Blanche and Stanley will be the final "kill" of the evening as Stanley devours the last of his doubters.

If you are involved in a production of *Streetcar* in which Blanche (as many believe) is the main character, your premise may be "Tirelessly maintaining illusion leads to the ugly truth." Your lighting designer will spend hours trying to illuminate the set bouncing the light off of mirrors and your makeup designer will have a field day with the ruins of Blanche's attempt at makeup. The climax will occur when Stanley rips the shade off of a lamp, revealing Blanche's naked face.

Some hints to help you in your premise include thorough examination of the title, characters' names, or repeated motifs in the show. (Notice: NONE of the above premises concentrate on the idea of giving into the carnal "desire" of each character.)

As you can see, we left the author's original intent out of this equation because we are seeking the analysis unique to this production. Sometimes the playwright may hit on the same premise, but what makes the me alive is the production of the elements the playwright pro-)f course, you will later examine the playwright's original intent, this early stage, why not try to create *your* interpretation? The exclusion of the author is controversial, so as always, adapt the process described above for your own purposes.

The premise you create must remain malleable as your collaborators are brought into the next phase of script analysis. Your premise is only a way for you to collect your analysis of the play before you work with others. This premise is important for stimulating discussion among designers, directors, and, later, the actors. By having this solid base, everything else easily falls into place. It will assist you in your next phase of script analysis. Please don't be married to your premise, though. A designer or director may disagree. But, as always, you are proving yourself as an intelligent and active collaborator.

> *R*: I refer to the premise in every step of the production. The actors are usu-
> ally sick of my constant harping on the premise when we make blocking
> decisions or character choices.
> *K*: I use the premise as the starting point and rely on it when I have ques-
> tions, but quite often put it aside as I begin to design. It is always there,
> but I find a production becomes more than just that sentence.

Since this chapter is solely about script analysis, we will postpone a discussion of background research on the play. Let us look now at additional readings of the script and your personal reactions to them.

Creating the Production's Style

The main purpose of the upcoming few readings of the script by the director is to begin to set the parameters for the production. The director's further examination may reveal a deeper premise. You as director must ask yourself many questions related to the show's design during this phase. These questions need not always have answers. That's the job of the designer.

In this phase the director will also define the style of the piece in relation to design. This does not mean you will decide on colors and specific elements, but the basic "feel" of the show. The "look" is up to the designer.

Some questions may include

- What is the play's genre?
- What is the location?
- What is the time of year—down to day and month?
- What is the ONE thing you want the audience to take way?
- What element resonates with you?

- What are your sensory reactions to this show?
- What is the most important element to you?
- What is the motivation for each of the characters?
- What do you think this production needs?
- What is the one thing VITAL to this production?

Your reactions to these questions might include finding evocative artwork or iconic imagery that helps you define this production and its style. Find music of the time or bring in aromatic elements that you think smell like the show. This sensual approach can offer a designer a myriad of choices. Be careful in this phase not to find an image and say, "That's *exactly* what I want!"

Inevitably, you will answer only some of these questions. Your job as director is not to be so specific that you have designed the entire production. But unless you answer enough questions, you risk being so vague that you offer your designers nothing.

Having completed this phase of script analysis in conjunction with background research on the play (covered in part I, chapter 3), the director is ready to enter the first design meetings armed with enough information to inspire.

Your analysis of the script will continue throughout the rest of the production. After reading for the design elements, the director reads it numerous times to decipher the characters' main objectives, to divide the script into units, and more. But for the purposes of this book, you are ready to collaborate.

The Designer's Process

The designer's process is similar. As a designer who has analyzed the script and has a clear premise (which may alter during subsequent readings), you will next read the script for practical environmental clues.

Some questions when reading might include

- When and where does the play take place?
- What is the timeframe in which all action occurs? — *Key in SF*
- What is the size of the world in relation to the story?
- Is this a real place to me or an expressive one?
- How many doors are there?
- Is there a need for lighting practicals?
- Where should they sit and how?
- Does the script demand special effects?
- How can I support the movement, the look, the intensity of the script?

Research will solve some of these questions. Designers must know the time in which the play was written, the time in which it was set, and understand the time in which we are living (why is it relevant now?). The process for the designer is never-ending. And it is just as important to find one hundred not-quite-right images as it is to find the *one* right image.

First, it is important for the designer to provide the director with imagery that communicates his own feeling of the show. Don't worry about the details at this point. Find your own pictorial interpretation of the script's true essence.

The same is true when talking about color. It is impossible to talk about the color blue. Bring in a thousand variations of blue. When the "right" blue is chosen, the rest of the spectrum is defined.

Preparing for Production Meetings

Now you, as designer or director, are prepared with the resources necessary to enter your first production meeting.

In a perfect world, your first design meeting should proceed as thus:

> *Director*: Well, what are your reactions to the script?
> *Designer*: I loved the part when . . .
> *Designer*: This part made me feel . . .
> *Designer*: I was reminded of . . .

Your job as director in this first design meeting is to create an atmosphere of collaboration. You are the moderator. Listen to the opinions of the designers and let them affect your decisions. You must guide the discussion if you feel it diverts too far from the premise, but you must also allow your designers to offer their experiences.

> *R*: Kirk and I had a design meeting about Lillian Hellman's *The Children's Hour* and Robert Anderson's *Tea and Sympathy*, to be performed in repertory. He was struck by one sentence in the final passage of *The Children's Hour* where the character is described as distractedly waving her hand to another character. He kept asking, "What does Karen's hand gesture as described by Hellman over fifty years ago mean?" This did not even enter my radar. I had read it, but this moment was not nearly as significant to me. The ensuing discussion between the designers and myself led me to adapt the direction of the end of the show.
>
> *K*: In the same evening when discussing the script analysis of *Tea and Sympathy*, I asked Rob who the main character was, and he said it was Laura. In discussing the play's inciting incident, which Rob believed fell in the hands of another character, I suggested that the inciting incident must support

the main character's journey if the show is to be interpreted in tł
He agreed with me and changed the inciting incident.

As designer, your role in this perfect collaboration will be to dis
experience of reading the play. You might employ the "director toucn
test." Present the director with several pictures of the sunset suggested in
the playwright's stage directions. The one the director touches and pulls
closest is the keeper. By using this method, you have now made a visual
connection on which to build your design.

Taking your cues from the director in terms of locale and time period,
you will hone your design decisions. Though you imagined a production in
which Mitch from *Streetcar* broke your heart, you must react to a director
who has envisioned the city of New Orleans to be the main character. You
may adjust your initial responses by trying to meld those qualities that
you find important in Mitch that align themselves with what the director
finds important with New Orleans (Mitch embodies the cultural values
unique to the city of New Orleans).

Identifying Collaboration Obstacles

Beware of conditions that may hinder collaboration:

*One issue
is whe
director isn't
defined —
Richard*

- <u>Inadequate budgets and short schedules often drive each decision.</u>
- Some directors may not be able to define for the designers a clear concept of the time and place in which the production is to be set.
- Some designers may not be able to relate their own—much less aug-
  ment the director's—vision.
- It is difficult in your short time together to truly trust your
  collaborators.
- Some directors, rather than inspiring, simply employ designers as
  facilitators and imitators.

R: I worked with someone who seemed to look to me for every design decision. I don't want a designer who makes me be a designer. I want someone to collaborate with me. At the same time, I *do* want a designer to
   react to my very specific goals for a production . . . but in his own way. Is
   that an impossible dream?

K: I worked with a director who doesn't respond to me at all. Being alone in
   this part of the process makes me somewhat neurotic. Is this silence a sign
   of his lack of interest in the design? Is it a bad time for him to communicate? Or is it just an acknowledgment that I responded well in the last
   production meeting? No matter what the answer, I find myself spending
   twice as much time as I rethink every decision I have made to this point.

Adjusting to a New Process

If you encounter the director who starts a production meeting with little or no analysis, you have our permission to jump to the fore with yours. Feel free to offer your opinions on script analysis and dramatic structure and observe how all will follow your lead. Congratulations on being the lead on this production. You may have inherited more than you bargained for as everyone comes to you for answers—a great position to claim, if you have the time.

> *You*: But Rob and Kirk, I have a director/designer who does not work in the way you imagine.
>
> *Us*: Too bad. We also have non-collaborators on every project! First, try to relate to your team in the way we describe. If that does not work, then use this as a learning process and do your job to the best and most professional level possible.
>
> *You*: I hate everyone's ideas!
>
> *Us*: There must be something exciting in their ideas. Another's vision, an image, or a tossed-off comment can pave the way to a productive process if you are open to collaboration.
>
> *You*: I don't like script analysis.
>
> *Us*: That is what you have been assigned/hired to do: analyze a script and bring it to life. You may not have found *your* version of exciting script analysis, but if you want to design or direct, you will analyze.
>
> *You*: I really like your ideas, but they never seem to work out the way you express them.
>
> *Us*: We know. We are writing about a perfect production process. It never works out the way you imagine or we theorize. There are too many variables. Your job is to adapt to the situation and maintain your collaboration goals at a level you can be proud of.

Using the Premise Creatively

- "How does good script analysis translate into a design?"
- "Does a premise really matter?"
- "Can't you just give us some hard-and-fast rules about how this works?"

These are all valid questions. The answers are as varied as our readers' points of view.

What is always true is that each production is unique because the collaborators merge their individual creativity. As the following illustrates, agreeing on a premise helps focus that creativity—and the production.

> *R*: When I directed a production of Lanford Wilson's *The Hot L Baltimore*, the lighting designer cued the sign of the decrepit hotel to blink at the inciting incident and every time the action would rise. This culminated in a blinking flurry at the climax. While this sledgehammer approach

to designing for script analysis may seem obvious, the sign was utilized in such a way that the audience understood the structure subliminally. Their comprehension of the sign's blinking was secondary to their experience of the lives of the characters in the show.

R: When I directed *The Rocky Horror Show* I created the premise that "unbridled experimentation leads to judgment." I had Frank N. Furter destroyed by Bibles with lasers in them held by the aliens Riff Raff and Magenta. The premise (in all its irony) led to a production where each element enhanced this hedonistic world of Frank versus the uptight and misguided moralists, Riff Raff and Magenta. The premise makes a production cohesive.

K: I was designing an operetta of Johann Strauss Jr.'s *The Queen's Lace Handkerchief*, in which the script had not yet been transcribed from the mid-1800s. I had a synopsis, but no script. Talk about improvising!

K: I was designing a new work by James Stuart and Robert Ward entitled *A Friend of Napoleon*. The design had been approved and drafting was finished. Then the director/librettist passed away two months prior to rehearsal. The artistic director of the company proceeded to take over as director. We tried to retain the original design for this new director's interpretation in an effort to maintain the creator's vision, but eventually had to change. The show maintained most of the original design, but ultimately the design evolved with the new collaborators.

No matter what your analysis, a production may incorporate your ideas seamlessly IF you are willing to fully support them. Your mission as director or designer is to complete your analysis of the script in conjunction with your collaborators to the fullest extent. Forget any distractions or blame, and simply work to your potential. And beyond!

As director, your job is to guide the production's focus (and subsequently the actors and audience) to the message you want to express.

As designer, your job is to enhance the production's focus and guide the actors and audience to the visceral experience you want to provide.

It is as simple as that . . . and as difficult. No matter your part in the collaboration, your duty is to serve the play.

CHAPTER 3

RESEARCH METHODS FOR COLLABORATION

*T*horoughly *researching a* play *is impossible.* There are always details you cannot even begin to tackle. Beside the imminent demand of production schedules, there are just too many elements that must be addressed. Even if you are lucky enough to have the assistance of a dramaturg, you can never cover everything.

Thoroughly researching your production *is possible.* Your job is to attain a general knowledge of the main ingredients of the show and then spend your time delving into the smaller elements that affect you most.

You will want to funnel your energies into practical, production-based research. By truly listening to your collaborators, you may save countless hours exploring elements that are not essential for the production at hand.

This chapter will build on the elements of vocabulary and script analysis covered earlier in the book (chapters 1 and 2) to help you create a collaborative process of research most valuable to all involved.

This chapter will show you how to

- use script analysis and vocabulary to assist in your research;
- research the production, not just the show;
- research as a director for collaboration; and
- research as a designer for collaboration.

Researching as designer or director brings up many questions: Where should I start? How will I know when I am done? How much is too much? How little is too little? What about the hours spent going down a seemingly wasted research path? Do I really need to know *all* of that? Most important, can I collaboratively research?

We hope to offer some insight to at least mollify your research paranoia.

Refining Internet Research

Most research now begins on the Internet. The ease and breadth of its knowledge offers you a wide-ranging sea of information over which to cast your net. The Internet offers you images, articles, and musings on nearly every subject you require—and many you don't.

The exercise of refining your search on the Internet helps you refine your future research goals in other media. For example, when Rob researched *The Children's Hour*, the very general phrase "lesbians in 1930" in his search engine yielded several not-very-compelling links. But among the interactive journals and sexual merchandise was a rare photo album of a lesbian couple chronicling their lives from 1930 to 1939. Although the images are unavailable on the Internet, the link pointed the way to a published book, a great aid for the director or costume designer.

> R: Stupid Confession #1: When initially researching *Tea and Sympathy*, I had no idea the title was a common phrase in use long before the play was written. It wasn't until I saw the numerous matches on the Internet that it dawned on me there may be more here than I initially thought.
> K: Stupid Confession #2: The first time I designed *Twelfth Night*, I searched and searched for images of Illyria. I mean, it sounded like a real place and actually was according to an encyclopedia: located in the upper Balkan Peninsula in the tenth century BC. Good luck finding photos of that. Who would have thought William Shakespeare would have done such a thing?

Distilling Further Research

Before you rush to follow the leads you've gathered, refer to our first chapter on vocabulary. This is an essential element in honing your research. If a director does not care about the line of your period costumes but rather wants them "to look pretty," you may limit your pattern searches and simply provide examples of what you can both agree on historically and aesthetically.

More vocabulary-based refinement may occur in all production discussions. Listen again to the keywords related to the description of the world of the play. In a discussion of Neil Simon's *Barefoot in the Park*, the director may want to place the production in present-day New York. All of your research on 1960s modular furniture is for naught. Or is it? You may certainly discuss your findings with the director, who may say, "Oh, how retro! I love it!" or "No, I want to avoid any clues that this play is dated!" Your past research will help you refine your new search for non-retro, modern furniture.

Script analysis is another great clue to assist in making your research more streamlined. Through discussion of premise and structure, you may sift and discard countless unnecessary details that you find useless in the upcoming production of *Romeo and Juliet*, set in the year 2525. You may spend less time on the "cords" brought in by Juliet's nurse (since this is played by a hologram) and spend more time researching the swo to kill Mercutio—a 2525 laser.

It seems as if we are advocating the easy way out. But selective research is anything but easy! You certainly must understand the historical context of the play as originally intended so that you can support the current interpretation in an informed way. Identifying the "wrong" research is important to you as a director or designer because you have to know what doesn't work for your current production before you can identify what does.

Through vocabulary and the clues of script analysis, your research has now become more productive and—dare we say it?—practical. You are now prepared to head to your local library, bookstores, and the newspaper's archives to really delve into your work.

Sharing Research

With your new streamlined research approach, you will run across imagery or information that will assist the director in discussions with the actors or other designers. By all means, *share* it with your production team.

Not only will that information be useful in honing the production team's discussions, but it also helps you see if you are on the right track. By finding imagery that may inspire another, you are again preparing to work collaboratively. By sharing research, colleagues truly create a collaboratively, trusting atmosphere.

There will be a few designers or directors who may feel you have overstepped your bounds by offering this imagery. As always, use respectful vocabulary to ensure they understand your intentions.

Since this book is about the collaborative—rather than the creative—process, we do not seek to make you a great director or designer (there are many excellent books on that subject). Instead, we offer strategies to become the best collaborator you can be within the production team.

Researching as a Director   — I felt this w/ the Amelia Project

As a collaborative director, you certainly must have the broadest range of research at your command to assist in all areas of the production. You must understand the play, the production, and your designers to make your vision understood.

Directors, for the most part, do not seek *actual* imagery to be used in the production. This limits designers. Rather, find things that evoke the mood or feeling you want. By selecting appropriate imagery, you will inspire rather than constrain.

R: I must admit that this concept is sometimes really difficult for me. It's hard for me to put aside certain concrete mental images after analyzing the script. Your research imagery just keeps pushing you into a design rather than an inspiration. I may talk like a collaborator here, but I often know what I want going into the first production meeting. Some shows, it just happens.

K: For the most part, I long for visual imagery from my director. I admit that I don't like ground plans thrown at me, but, if I cannot connect with the director through discussion alone, I ask for an artist or a specific piece of art that they feel inspires the production. I worked on a production of Theresa Rebeck's *Bad Dates*. After the director referenced a specific painter, I selected one painting from the artist's works in my attempt to understand the director's personal connection to this play. In the end, I used the portrait on the actual set.

A director's duty is to nudge the designers so that they may transform the production from the idea phase into reality. By presenting visual, sensual, and aural research to each member of the team, a director makes defining the world of the play a comfortable and shared process. The director has to find any means to address all aspects of the production—from overall style to details.

The research we are describing throughout this chapter has very little to do with the historical and more to do with the imaginative. Certainly, historical research affects the collaboration. But for our purposes, we will concentrate on the intangible and evocative elements.

Let's imagine you are a director preparing for a production of Tennessee Williams' *Summer and Smoke*. You love this play! You relate to the main character's (Alma's) internal struggle to express her true sexual desires. You see many ways the show could go, but you really think that the main issue lies in Alma's repressed sexual desires. You need to find imagery to inspire your designers. You want them to understand and, on some level, relate to your vision. So you

- go to paintings and photographs of houses of Mississippi. These all seem *too* literal;
- find pictures in another book of swamps that make you think of the figurative morass in which Alma finds herself;
- see in another the kudzu choking the trees—Alma again!
- find on the Internet a famous photograph of Flannery O'Connor sitting on her back porch next to a peacock. This image of a frail woman next to this proud bird is exactly the kind of clinging to the accepted way of life that holds Alma back;
- listen to some Sousa marches played by a community band that are just awful and exactly the kind of pathetic celebration that begins the play;

- find a tattered lace handkerchief of your mother's that has been unused for years but is still crisply starched;
- keep seeing in your head different bars when you think of the show... Prison bars? Cages? Wrought iron?
- hear the calming waters of a fountain as you sit outside reading the play again and realize how antithetical this sound is to the storming sea inside Alma.

All of these images and thoughts help you prepare to collaborate with the designers. Hopefully one of these disparate elements will inspire the designers to create the world of the play.

Researching as a Designer

To be the best collaborative designer, you must certainly have the creative talents to bring research to life. The designer must, through guided research, create a part of the universe of the production in conjunction with all else. Your great research and resulting design may mean nothing if your fellow designers aren't similarly inspired.

Designers must also find unique and interesting ways to visually represent this particular production. All must find images to make the production cohesive under the umbrella of the premise.

Recycling Research

Imagine that you are a lighting designer. Immediately upon notification about one month ago that you were hired/assigned to design Tennessee Williams' *Summer and Smoke*, you read and reread the script. Your script analysis led you to discover the clever word games of the character's names and a major theme of religious persecution, which really spoke to you. Then you started collecting some imagery for the director which communicates your reaction to the script. You found historical paintings of the Spanish Inquisition with its smoke-filled chambers, photographs of shadowy halls of secret religious societies, and pictures in magazines of the harsh sun scorching fanatic Islamic fundamentalists.

You attend your first production meeting and discuss the script. Everyone seems to agree on the dramatic structure and the playwright's original intent. The director then reveals her premise and overview of this production. The director ends the meeting by saying:

> *Summer and Smoke* is about those two words only. I refer you to Alma's speech in which she mentions the title—that part of her died last summer, suffocated by the smoke of her desires. In our production, Alma will

be loosened by the heat of the summer and choked by the smoke of the smoldering desire of John. The premise of the production is "un-fanned smoldering passion leads to dying embers." The religious elements in the script don't interest me other than as peripheral obstacles to suppress desire.

You didn't even get a chance to show your initial research. The director wasn't even interested in YOUR interpretation of the script. All of your research on the lighting was wasted. Or was it?

Why not share your religious iconography at that first production meeting? Why not offer your analysis of religious persecution to the director? Why not share your vision of the production? If the rest of the team is unreceptive then preface your presentation with the appropriate, assuring vocabulary to make sure that all understand that you are willing to collaborate:

> I know now that we are concentrating on the suppression of desire in this production. So ignore the rack in this painting of the Inquisition, and just give me an idea of what you think about the way the torches have made this room smoky. Is that the kind of haze you are looking for in the opening scene at the fireworks?

Even though the subject of your research may no longer be relevant, some details within may point your way to more guided research. Hopefully you can find positive elements within the imagery you presented that interests you both.

At this point it may be helpful to confine your research to pertinent imagery of Southern summers, smoky plantations, and hazy days on the Gulf Coast. This research is broad enough to help you find the look for the production and specific enough to the text so that your time gathering it will most certainly not be wasted.

Further Research

As designer, immediately upon leaving that meeting your research begins a new phase. Before hitting the library or photography magazines, take a moment to explore the best resource—yourself.

The best place to start your research is with yourself. Armed with the concept for your production, begin this phase of research by asking yourself guiding questions: "Was there a time I felt what the director described? What elements can I recall from that part of my life to make me personally connected to Alma's plight? Have I been to Mississippi? Have I felt that desire choking within me? What does the act of choking literally feel like? What do I have within me that remains un-fanned?" Once you have tapped your personal connection to the premise you may begin to research other elements. If each collaborator were to follow this process, a richer communal production experience could take place.

As lighting designer, you would refer to your notes from the meeting and examine how you may contribute to the premise. You may also confront the lighting issues the director was most worried about. Those damn fireworks and that oppressive humidity kept being mentioned. Maybe you should include a hazer or support a downstage scrim for the lights to really get the effect she mentioned. How can you research the best elements for your production?

You'll call your friend who lives in Mississippi or you will research the weather or search for photos that best represent that oppressive feeling necessary for the atmosphere of this production—hazy, humid days where everything hangs trapped and suspended in the air. These images resonate with you. "Trapped by oppressive heat. I know what the director wants!" Having collected your imagery you stand proudly ready to collaborate at the next meeting.

Upon entering the next meeting, you proudly present your research and the director loves it. No designing has occurred. It is purely a meeting to share research and personal experience. The director is thrilled that you have found the right color palette and impression for the show. The costume designer on the other hand really hates the gel colors you are using because of the dresses being used from costumes in storage (which are perfectly in tune with his own take on the premise), which will look orange underneath your gels. "It simply won't work." The set designer is afraid that the scrim you mentioned using for the lighting will negatively impact the detailing of the molding that she has thoroughly researched (and is perfectly in tune with the premise). What to do? Collaborate!

Balancing Research

As you can see, the costume designer has taken the lead in the collaboration. This designer has already chosen exact elements around which everyone must work. The meeting that was called to simply share research and to see if all are on the same page has now turned into a reactionary response that throws off the balance of the collaboration.

Not to worry, things like this happen all of the time. You have the necessary vocabulary to make your opinion known without being destructive or defensive, you have the text and premise to support your work and finally you have your research. This is the moment when the collaboration on your end truly begins. The territorial battle for design integrity is moderated by the director:

> *Director*: Nice work everyone. How can we make all of this research work cohesively?
>
> *Lighting Designer*: I would like to have a light lab for the costumes to see what possible gels I can change if we are using those dresses.
>
> *Costume Designer*: I am afraid for budget reasons we have to use them, but I could possibly alter them in some way…
>
> *Director*: Why not use some overlay fabric that would get the shimmer I see in the lighting imagery.
>
> *Scene Designer*: I could also mimic that in the rectory windows.
>
> *Director*: I like it!

Although the above is considered collaboration, you may go to the next level and begin a respectful discussion about how each of the designers came to the choices they love and perhaps brainstorm together on ways to find new potential possibilities that will work collectively. Although this kind of collaboration is more difficult to monitor, it sometimes pushes each collaborator to a higher level. This process is truly collaborative.

> *Director*: All of these elements are individually right on the money. How can we make this show unified?
>
> *Costume Designer*: What I like about these dresses is that they are so restrictive and yet there is a hint of sexuality peeking through.
>
> *Lighting Designer*: Yes, I see what you are saying. I like how you have Alma's transformation so clearly delineated and that final costume completes it beautifully.
>
> *Scene Designer*: I also like the photos you found, which recall the restricted Alma and conversely the more sexual Alma.
>
> *Lighting Designer*: I am on the same page as I think that the costumes for Alma really establish this very specific journey and I can definitely compliment that.
>
> *Director*: But that final scene at the train station really needs to reveal an entirely different feel than everything else. Alma is destroyed, drugged, and coping with her sexual desire by performing the act she pushed away for so long.
>
> *Scene Designer*: I really want isolation here. How can we show the deterioration of the town and more specifically Alma?
>
> *Lighting Designer*: I can help there. It is already in the costumes, I think I can amplify the starkness of the isolation that is in your set image of the lonely Hopper painting.…Do we like this?
>
> *All*: Absolutely!

The difference in the dialogue above is that the collaborators are speaking to each other. They are speaking beyond the practical, covering the play itself and more importantly this production. Ownership and agreement in each element can create a cohesive and deeper whole.

If you find that you are unable to get your point across to designer or director—by all means educate yourself. Audit a directing or acting class to assist in speaking to a director concerned mainly with the actors. Sit through a technical rehearsal or a production meeting to observe designers in action if you feel ill at ease when trying to share your director's vision.

> *R*: I consider myself weakest when speaking to lighting designers. I had the least training when it comes to this element and I think it so ephemeral and intangible that I feel I never have the correct vocabulary. I asked graduate students to take me through mock lighting labs and tell me how a director could speak and collaborate with them. It seems I wasn't that far off, but my insecurities made me question every note I gave.

> *K*: I consider myself weakest when working with a director whose vision I do not agree with. I do my part and present my ideas and draw from the director ideas that I can latch onto. I endeavor to find the right image or concept that can bridge the gap between the different productions in our heads. It is not my desire to alter the concept of the director or redefine the premise, but to connect on some level. If we cannot find this link in the images themselves than it must come from other elements—descriptives within the research images, the room in which we are meeting, the tree outside the window, or merely in dialogue. Whatever works...works.

Implementing Research

You now have a clear understanding of the production and the director has presented a clear vision. You have been inspired by the images that you and your collaborators have shared. You know the style, the feel, and the premise. Now, you must do what your title implies—design.

Putting your research into practice is the next step and as designer this is the most terrifying moment of the entire endeavor. The budget, the schmoozing, the meetings all come down to this moment! Putting pen to paper (or finger to mouse).

CHAPTER 4

PRESENTING THE COLLABORATION

You are a designer. You have listened to your collaborators. You have done your research. You have put pen to paper and you have designed. You designed with the audience's reaction in mind. You designed the premise. You designed the climax. You have designed with the needs of others in mind while allowing your artistic abilities to create bolder, better plans for your contribution. Now, how do you best present them?

In this chapter we will explore variations that you may encounter when presenting your work as designer or director. This chapter aims to assist you in moments of highest insecurity for all involved: the day you share your work with your production team.

This chapter will show you how to

- present your design;
- respond as collaborator; and
- respond to criticism.

Initial Presentations of Designs

"I would like to show you what I came up with."

Some may think you should let the work speak for itself. We think it is best to take your time and walk the collaborators through your design slowly. This does not mean that you should describe EVERY decision and stroke of genius but rather a thorough explanation addressing the needs of the others and explaining the function of your product.

FOR EXAMPLE: Kirk designed Theresa Rebeck's *Bad Dates* (see preliminary digital rendering, figure 1)

Here is a basic outline of Kirk's presentation:

- Here is Haley's world.

I always like to recap some "general" ideas as I start into the presentation.

Figure 1 *Bad Dates* **by Theresa Rebeck, preliminary rendering**
Source: Directed by Kristine Thatcher, Scene Design by Kirk Domer.

- The director wanted a world of old and new, a world of solid and transparency, a world of mother and daughter. (With all of this in mind—it is a one-woman show where Haley speaks directly to the audience.) The design is relatively simple to frame Haley's story. We are able to see this bedroom as well as the world outside because Haley needs to prepare for dates in the bathroom and speak to her daughter down the hall.
- The bed is central and most important in the scenery to draw attention to Haley's sexuality and to enhance the "new" aspects of her life. She has spent money here to make herself feel better as a person, empowered as a woman, and successful in the fact that she now has the bed that she always wanted.
- I have provided islands throughout. A seating area—to try on the thousands of shoes she goes through during the show. A make-up table—to prepare for these dates, in another area to show she is utilizing every inch of this apartment. A window seat area—to give reference to the outside forces that affect her daily life. An "open" hallway to her daughter's room—to see the important connection she has.

The daughter who is never seen in the play was an important element for this director. She loved a painting by Mary Cassatt and thought it spoke volumes about the lead's relationship with her daughter.

- I used a print of the painting in the area where her daughter is as both an element for director and actress to connect with.
- There is a full-length mirror on the inside of the SR closet door as discussed earlier.
- The bathroom and "quick-change" area are behind the velvet headboard for privacy and to make the wardrobe crew's life easier. Even though the walls throughout are opaque, the solid headboard wall will be helpful for all involved.
- I wanted to skew the direction of the flat set in this awkward performance space so that the lighting designer could get some interesting angles and help me avoid the flatness of the design.
- Finally "my favorite part of the design" the dress form—so Haley can dress and undress the "mannequin." Not only could it supply great business for the actress but it populates the stage in this one-woman show. Haley can actually dialogue with this new character transforming the dummy into people only spoken about.
- That is it—Haley's world.

Directors and collaborators: DON'T interrupt constantly, jump the gun, or point out flaws. Let your presenter speak. Most of your questions will be answered by the time the presentation is finished. Your interruptions may throw the presentation off track or cause some unneeded friction.

Once the design has been presented, it is usually customary for the director to respond first. As director, before you speak, consider what has been presented before you. Some questions to ask yourself as the presentation occurs: Did my collaborator understand and adapt what I was talking about? Did my collaborator take into account my needs as well as the other collaborators? What do I like about this? Where am I confused? Where did my failure or success of communication enter into this design?

You might take notes as the presentation occurs so that you are sure to remember your thoughts. If your collaborator has been working as hard as you have, you should certainly give the person the benefit of a carefully measured response. Always begin with the positives of your reaction. This may sound elementary or condescending, but the moment a director sees the preliminary work is both exhilarating and terrifying—so anything may escape. It is also the first true test of the director's communication skills. If the design team is all off the mark, the director may have some serious rethinking to do involving the way she communicates her concept.

Let's imagine a director just saw a design in very early sketches that was so far from anything imagined:

Director: I am confused as to where you were heading with the overall feel of the set. It seems very heavy when I thought we looked at ethereal images and talked about open vistas.
Designer: Yes, well, I tried that and I thought about how it might be more powerful to just see glimpses of the sky behind that huge wall.
Director: Well, I really think this set would work well for the dramatic version of this production. This pastoral comedy can't really support the weight of this stone wall. It might have been here at a time before this play began. What if there were remnants of this wall that have been used now as places to worship the sky and countryside.
Designer: What do you mean?
Director: Is it okay if I draw a little here or move your model around?
Designer: Sure.

R: I always feel like I am destroying a piece of art when I start moving or adapting a model because all those little pieces take forever to make. Ugh! I feel terrible.
K: I love to rip things apart or scribble with the director. Yes, it took a long time to create and build a model or paint a rendering, but, if you treat it like "displayable" art and not like "idea" art, you can really stunt the creative process. Take scissors to the model and maybe an ink pen to those drawings to show the director that you can adapt and explore to make the design work for the production.

An amenable designer can begin manipulating the idea to something workable. Remember, it is perfectly fine to say let's go back to the drawing board (literally!) if you can't make these initial ideas work. If that is the case, it is then imperative that the director clarify the needs of the production so as not to frustrate the designers.

Solving Issues

Let's imagine you are a designer, and a director obviously does not appreciate what you have done . . .

Director: Um . . . it's different.
Designer: Different in a good way?
Director: In some ways.
Designer: What works for you in here and what do you find problematic?
Director: The entire flow of this section leaves me no traffic patterns, no good entrances, and no levels! How can I have thirty people on stage in one clump all on the same level? It's like the unemployment line!

Designer: Well, I was going with your images of the plains and how
into the flatness and ordinary quality of these people'
you remember that Andrew Wyeth painting you brou~
loved the expansive flatness . . .

Director: LEVELS!

Designer: How about a raked platform that enhances the idea of the flatness
and also will give you levels. We can combine both ideas?

Director: Genius!

Director's Response

I found this positive language [support] /talked well

Your colleague has just presented a design. Your job as fellow designer is
to wait until the director finishes responding before you begin to ask any
questions you may have regarding your colleague's creation. Most of your
questions will be answered in the presentation or the director response,
but, if not, remember to respect the integrity of your collaborator (whether
or not the designer seems to have missed a part of last week's discussion!).
Remember, your job is not to critique your collaborator's work but rather
to respond and build from the individual's contribution.

Let's imagine that you will have no place to hang lights on the preliminary design of a set. . . .

really liked . . .
"I like this . . ."
"can we
try this"

Director: I like it! I see nothing wrong with it! Anyone else?

You: Where does this sit in regard to the electrics? It
seems that I have no side-lighting opportunities in
order to "pop" the actors and enhance that trapped
feeling. I am confused by the height of the walls and
placement of the portals.

Designer: Well, I hadn't really thought you would want to do
side lights so I turned all of the portals sideways to
make the stage a complete box.

Director: Fascinating!

You: Yeah, it really gets the trapped quality from our initial research. But I do have to say that I will need to
have some side lighting to avoid the flatness of the
light and the shadow on the faces that could really be
an issue when we go into technical rehearsals.

Director: I hadn't thought of that.

You: If we just angled the portals ever so slightly we would
achieve the same effect plus I could easily enhance
the lighting and provide more depth. Plus one resid-
ual positive is that now there are entrances and exits
possibilities for the actors.

Director and Designer: Genius!

While again (as in all scenarios in this book) our dialogue veers into hyperbole, the essence of each sketch is real. The basic premise of "mutual respect leads to a healthy environment" holds true. But when simple politeness fails and fights ensue one must truly collaborate in the midst of a hostile environment. If one is unable to step back from "personal attacks" one must continue to champion a cause for their own conceptual ideas. Rather than being dragged down into the mire of name calling one must rise to the basic needs of the production. Any argument must end with the production's core remaining intact.

Production Meetings

If you have completed your presentation and received feedback, it is your responsibility to make sure what the expectations are for the next meeting. As director, make sure you have a clear idea of what you will see next time. As designer, make sure that you have enough information to spark a new round of creativity that addresses or alters any concerns that were discussed.

Maintaining Cohesion

In this moment of the collaboration there is one issue that you must address: do all elements agree? What does that mean? The lighting designer may have an idea for a set of gothic candelabras, which is fabulous. The set designer has the idea of shoji screens, which is elegant. And your costume designer has gone with black t-shirts and black pants, which is simple.

In this obvious example we see these separate decisions lack cohesion. Even though a team may get along swimmingly, their product must also work together. So, when a team has created a world where styles have been mixed without justification (other than "It is pretty") it can rarely be considered a successful collaboration. The measure for all discrepancies is the premise and the production. If an element troubles you, keep asking, "Does it serve the premise and the production?"

Your job in this presentation phase is to discover what elements of each collaborator's contribution works for the good of the production and how they work together. The unique elements of each design must work collectively. In this presentation do not leave the room without pinpointing the elements that work for each of you and including an explanation of WHY!

As you may know, these elements must be allowed to transform throughout the collaboration, but this presentation is the navigation session to help guide all to the final cohesive product.

Perhaps you like the set designer's rubble strewn street and are drawn to the evening gown sketches of the costume designer. Ask yourself, "How does the tension between these two ideas change the production?" Once you have answered that, ask yourself "Is that the production I want?"

  R: In a past production, a designer wanted to place cartooned, two-dimensional tables next to fluffy, three-dimensional chairs and realistic props. That can't be. Even within this own designer's area, the world did not agree.

  K: You need to ask yourself which is more important. Do I want 2-D with no life or fluffy realism? Both are good solutions, but which works better for the production and the director's concept? And ultimately which of these ideas are going to be utilized? Function must make an entrance into this scenario.

Presenting II and III

Each successive round of presentations to the team must always begin with a recap of what you heard from the last meeting. By reminding your collaborators what you felt your mission was at the end of the last meeting, you can focus the response to address your solutions. You may also assist the collaborators in helping focus your work if you misinterpreted any part of the last presentation response.

Once you have put forth your agenda in addressing revisions to the design, talk about your solutions and new ideas. This is the time for all to see your sensitivity to the collaborative agenda. Did you solve the lighting designer's issues with your new set angle? Did you adjust your palette in the costumes to compliment the set? Did you adjust your concept once the set designer pointed out the discrepancy in your premise? These important questions are all valid and usually posed internally by the other collaborators during this session.

Once you have adequately and concisely presented in these later rounds, it is possible to touch, tear, and mold your physical designs to make things work cohesively. Allow your work to be revised on the spot. The time wasted in fear of creating in front of others is outweighed by the proximity and time of all involved. Design in the room—the team certainly understands the problems with such pressure. Simply treat it as a sketch rather than a complete work of art.

If a designer's note has not been addressed adequately, offer more details. If a director is not pleased, ask for more guidance. If this trend of miscommunication or radically ineffective designs continues for more than two meetings, lock yourselves in a room until the design has begun to take shape that all can live with. The refinements may occur once the room is unlocked.

Presenting to the Actors

A production team usually presents its design one last time, to the cast of the show, usually at the first rehearsal. This moment is one our favorites. Although most actors usually coo over the general feel of the fabrics or scope of the set, there are several who would benefit from a deeper explanation of premise through design. By involving them in this process, it enriches the actors' idea of the style and importance of certain elements of the production—including the acting. Do not do a generic and cursory talk through of your design, allow the actors to hear some of your process. Allow them to see what ideas you discarded in order to define the environment. It makes you a better communicator and begins to ensure that everyone (including the actor) is working on the same production.

We have begun to turn our rehearsal rooms into "living museums" an idea espoused by Pamela Howard in her book *What Is Scenography?* to assist the actors in immersing themselves into the design and feel of the show. The research of designer and dramaturg, such as mood boards (discussed in part III of this book), cover the walls throughout rehearsals. The models and renderings are on display as practical reminders and further inspiration. The fabrics are available as tactile reminders of the costumes. The renderings of the design inspire the actor to assimilate into the designed world that is usually not introduced until tech week. It is a lovely gesture by the designers, a needed reminder for the actor, and an invaluable tool in creating a cohesive production.

Transforming Ideas

Following presentations comes the revision phase. This is an important and often hurried part of the collaboration. Time should be allotted for the design to breathe and transform. The designs should evolve. The sad part is most times it is simply not the case.

Most times the director says to the designer, "Just do it like this!"
Most times the designer says to the director, "Just tell me what you want!"
Both of those interjections are problematic.

CHAPTER 5

REVISING THE COLLABORATION

This chapter will show you how to

- overcome variables in collaborative issues;
- continue to collaborate through revision;
- maintain communication between collaborators;
- maintain the chain of command;
- redefine roles in the collaborative team;
- learn how to "let go" of your favorite elements; and
- complete a cohesive design.

You have presented your design to your fellow collaborators. Their responses were measured but positive. Your response to their work was measured but positive. You see some elements in everyone's work that presents larger issues, which could be problematic in the future. You see elements of your design that do not mingle with the elements of others' to create a cohesive production. This is exactly where you should be in the process.

The idea that the first presented designs will reveal a complete, cohesive production is rarely the case. In this chapter, we will help you discover the ways to revise your work in the most productive and efficient manner possible while still retaining your design and conceptual integrity. This sometimes furiously paced phase of the collaboration is one of the most unwieldy in terms of keeping the collaboration secure. This wrangling of divergent agendas and ideas during this time is difficult to say the least.

Negative Reactions

- "Well, he hated it!"
- "She made no sense when she told me about her concept!"
- "After the notes I got, why doesn't he just do it himself?"

The mood of the designer following presentations is usually despondent, defensive, or diffident. All three of these responses usually come out of a feeling that your work was not respected or understood. The care, attention, and time of the designer was brushed over by the team in a cursory response of a few minutes when you spent countless hours poring over the script and research plus the actual designing time.

- "Did she even HEAR what I was saying?"
- "Why would he even think of designing it that way?"
- "This is going to be so bad!"

The mood of the director following presentations is usually depressed, disturbed, or deserted. All three of these responses come out of a feeling that NO ONE but you understands what the show is about! The care, thought, and time in presenting the premise and the concept backed up by countless hours of research seems to have resulted in a hodgepodge of unrelated designs that will never work together.

At the best of times, the presentation phase is followed by an unending buzz and free-flow of ideas that cannot be stopped. Every sentence begins with the words "What if we." The ideas surge and ebb with refinements and new thoughts. Those are the moments that directors and designers yearn for. Those moments are rare. They cannot be fashioned—yes, even WITH the help of this book. Although the groundwork for such moments has been laid throughout, there are too many variables. We offer a vision and concepts to achieve cohesion among the many variables, but it is theory. This chapter deals with more of the practical.

Overcoming Negativity

As we write this book, our collaborations have not stopped. We are still working with other directors and designers on new production teams. We try to practice what we preach, but the introduction of reality into the theory proclaimed in these pages produces challenges. We will chronicle a few negative experiences and offer ways to overcome them:

> R: I was working on a production where the costume designer did not understand or agree with my choices for my imagery collected for the feel of the show. Although one of my favorite collaborators in the past, she had no idea how I could be leaning toward no walls or doors on a set, harshly stylized lighting, and neutral furniture in combination with realistic costumes. How did those elements jive? Because she was a big supporter of this book, she used its theories to prove her case about a cohesive production. To her these disparate choices added up to a MESS. My first response was, "Have you never seen a show like this?" When she said no, I asked my other collaborators if they also shared her feelings. It seems

they did and had not said anything. It was my job to get them on board OR rethink my plan. After serious consideration I decided to stick with the imagery we had all collected. I knew what I saw in my head—and short of drawing it, I had to find similar productions or images that combined these "odd" elements. I remembered pictures from Adrei Serban and Santo Loquasto's productions of Chekhov plays from the 1970s. I can still see the realistic yet symbolic baby carriage and the open walls with the real couch flanked by an endless row of cherry trees in the orchard. Once the designers saw these photos, they understood what I had been unable to express. The beauty of the photos combined all of my separate research into striking images that made them understand the design possibilities of the show...except for the costumer who replied, "I still don't get it, but I'll do it." Although I did not reach her, the history of our working relationship allowed her to go along with the idea. See photo of the final result (see figure 2a).

K: I was working with a director with whom I spent countless hours on the phone discussing the feelings, emotions, and imagery we found vital to the production. Then we spoke again and reaffirmed everything that was important in our first conversation, yet more words and more imagery continued to arise. Finally, I put my pen to paper, implementing every image we discussed using every feeling on my part to communicate the idea. The director hated it. The next conversation was short and sweet. He said, "These renderings aside, how can I communicate this to you better?" A new flurry of phone calls, this time reiterating the character development over and over again, which made me realize that this was the only consistent part of each and every phone call we had. Though we had worked together several times and enjoyed creating large and elaborate settings, this show was not about the scenic world. It was about creating a blank canvas for the characters to come to life. See photo of the final result (see figure 2b).

The realities described above are—more often than not—the norm. Divergence from our utopian collaborative experience is expected and accepted. Below are several possible scenarios and possible solutions once you have exhausted the obvious ones:

- A director who over-explains and bores you into a comatose design— Try to shock him into a new way of thinking with your design or ask only for concise written responses
- A director who has grandiose ideas that are impractical—Try simplifying the ideas into the essence of the show or offer budget realities to bring her down to earth
- A director who micromanages every detail down to the color of the character's underwear—Try to offer several choices and ask him to

Figure 2 (a) *Tea and Sympathy* by Robert Anderson, Michigan State
University (*Source*: Directed by Rob Roznowski, Scene Design by Kirk Domer,
Costume Design by Karen Kangas-Preston, Lighting Design by Shannon
Schweitzer). (b) *L'Enfant et les Sortilèges* by Maurice Ravel to a
libretto by Colette, Eastman School of Music (*Source*: Directed by Ted
Christopher, Scene Design by Kirk Domer, Costume Design by Nellica Rave,
Lighting Design by Nic Minetor).

point to one of your color samples or (avoiding a passive aggressive tone) simply offer him a description of your duties as you see them—and that includes designing the costumes with his opinions and not his artwork.

- A designer has no imagination and you want more fantastical elements in the design—Try defining what "fantastical" means to the designer by asking him to sketch a quick version of this play's world or ask him to name the three most important elements of his very leaden design and start molding and expanding them.
- A designer offers no research or does not contribute to the team—Try giving deadlines of what you need or keep asking her to respond to another's work during presentations.
- A designer (or director) just doesn't like you—Try getting through this experience with dignity by keeping calm and thorough in your work. Usually you'll get a thank you card on opening night at the very least.

Collaboration Outside of Full Meetings

Be careful not to leave people out...

As we mentioned, this stage of collaboration may be most cumbersome. Most of the frantic aspects of these revisions occur in private meetings. Design and production meetings may be set, but the real revisions occur between these meetings, often without other members of the production team present.

For some, once the meeting ends designers can finally talk with one another about issues without the director watching over them. The designers offer each other ideas or collaborative compliments. Designers can meet for lunch or discuss ideas quickly in the hall and share their research on a deeper level. Perhaps they can help clarify or even redefine what was meant when the director mentioned "minimal." Perhaps they can really just play off of one another with quick sketches. Perhaps they could bond over the frustrating way the director always starts his notes with a positive statement.

This approach is dangerous for several reasons. One designer is always going to be left out of the loop and it may compromise the trust of the team. When the designers who met come in to the next meeting with a decision they "love," their conference may throw off possible honest collaboration. And finally, a director may think that the designers are always complaining about him behind his back.

A director might like to chum up to a designer in the hall and redefine a concept. The director might also have been thinking about the question the designer raised and can't WAIT to discuss the breakthrough. The director might have understood what the designer was going for in that vetoed sketch, had an epiphany, and now loves it!

In this separate approach, the other designers feel left out of the process. The chance for honest and equal collaboration is now broken. The collaboration may be too heavily weighed toward one design area. All are possible reactions to these private meetings. All of these scenarios are natural, as well as dangerous.

The answer according to us? You must have the freedom to bounce ideas off each other in the hall, BUT you must make sure everyone is kept abreast. These moments of shared intimate collaboration are excellent, but all must be aware of these discoveries to keep everyone involved. Perhaps a daily email to keep everyone informed? Perhaps a private website to post updated designs?

Prioritizing Issues

Throughout revisions and the refinement process, it is quite easy for the collaboration to suffer due to time, frustration, and creative exhaustion. It is best to never simply settle for an easy way out or delay a decision for later. The settling on one element impacts every future decision of the production team. One element that is not cohesive or sub-par in the production causes a domino effect of lowered expectations and justification.

At the same time, a solution of delaying the decision until later in the process usually leads to making that decision when the other elements of the production should be taking precedence. When the blocking of the show moves to the fore of a director's agenda, the director does not like to continue to make decisions that could have been solved earlier. Conversely, a designer who is finishing up costume details does not want to go back and rethink that Act Two dress with previews starting soon. The alterations throw everyone's entire process off. Delaying the decision usually means you will still hurriedly make a decision at a later time. Unless this delay is used to gestate how this element can enhance a production or take the collaborative cohesion to a higher level, the solution is usually an afterthought.

> R: In a production I directed of *The Life* [chronicled in part II of this book] we delayed a decision for a major scene in the show. It was only one scene, but it was pivotal. The look of the rest of the show was detailed and harmonious—this element was not. The designer [Kirk] and I settled for something which neither of us were very proud. This decision was made during tech week when we were both exhausted. As you can see, we broke every rule above and suffered for it. Hopefully we also piqued your interest to read part II of the book.

K: In a production I designed of *Oklahoma!,* I continued to delay the treat-
ment of the floor. In this stadium seating auditorium the floor becomes
the backdrop. I waited to see the completed costume renderings. Fabric
had not yet been purchased because the director and costume designer
were having troubles choosing the color for Laurey's dress. I delayed
the painting until the last minute. As first dress rehearsal, Laurey's tur-
quoise dress drowned in my Grant Wood-style blue centerpiece of the
floor. She was hurriedly reblocked to avoid that part of the stage.

What decisions are reasonable to delay in this phase? What is a reasonable amount of time in this delay? There are several decisions that can
wait. These are called details and refinements. What cannot wait are the
major elements of the collaboration that are crucial to the premise and
concept. The revision phase is all about addressing the needs of others and
coming up with solutions. It does not need to be about the trim. Do not
move forward to discussion of details until everyone have agreed upon the
major elements.

TIME: If you got a late start on the collaboration and rehearsals
begin in a few short days with no finalized ground plan, there are several
options. First, meet everyday until rehearsals begin. Second, see how long
the director plans to spend on table work before the actors get up on their
feet—that may buy you some time. Third, attend a few rehearsals and hear
the play read, see the actors move, and watch them do some improvisa-
tions to inspire a ground plan, built specifically for all involved.

FRUSTRATION: You can't imagine what a director wants when he
proclaims that he despises seeing gobos, and yet he wants texture in the
lights. All of the gobos and images you keep showing are rejected. The
director has seen every gobo in every catalogue utilized in past productions and asks for new ideas. You have no answers. You believe that the
director is being unreasonable. A few options: (1) Present the director with
several originally designed gobos; (2) Provide alternative ways to simulate
texture on stage with the configuration of you shutter cut; (3) Maybe texture in its shadowed form is not the texture you are looking for—use the
scenery to accomplish this.

EXHAUSTION: You are working on three shows right now and you
really don't want to spend the time revising your ideas for Mayor Shinn's
costume. You know you have one that will work in stock. It is the same
blue suit used in *Fiorello!* the year before. One mayor's suit is pretty much
like any other. But before you make that decision out of sleep deprivation,
remember your audience saw the same suit last year. Also, your corner-cut-
ting sets a bad example to those altering Mayor Shinn's costume. They may
see your exhausted easy answer and use a similar ethic when hemming his
pants.

Maintaining the Chain of Command

During this revision phase, the chain of command established in early meetings may become blurred. Sometimes (as we suggested in chapter 1) this is a good thing. We do want to aim for that shared easy exchange of ideas between areas, but more often than not, in revisions, the selfish nature of the work can rear its ugly head.

You become territorial for several reasons. You want to make sure your contribution stands out. You want your element to remain just that, your element—after all you thought of it. You don't trust that the director has a plan for this production, and *you* will handle the agenda in future meetings in order to create the best production. You have already worked five times harder than the rest, and your work is done. You simply don't think the others have designed with your work in mind.

This territorial approach is antithetical to everything this book is about. By selectively "collaborating" you offer no room for the personal and creative growth that could spring up from challenges that other team members can inspire. By maintaining a less territorial approach and respecting the roles of others established in earlier meetings you may inspire and be inspired.

Altering the Design

We have already covered the importance of referring to the premise for every decision. We have also made sure that you are always aware of your primary responsibility to the audience and their experience viewing the production. One more element to help you problem-solve appears during revisions—the switch.

It is now time to switch roles. You have now seen every one of your collaborators in action. You have heard and understood their design "vocabulary." You have seen their research take shape. Now, it is time to "become" them when revising your work. True collaboration occurs when you begin to shape and alter your revisions with others in mind. By recasting yourself in the new role of designer or director, you may solve many frustrations sometimes caused during revision.

In your office or studio, imagine actual dialogued responses to your solutions. Imagine what this director will say. Imagine what a fellow designer will say—not only the negative, but also the positive. By remembering their presentation style and research imagery you will save countless hours of dead-end tangents. By understanding your collaborators you can aim for, and achieve, true collaboration.

Caution! This process should not take you into second-guessing every idea with a "She'll hate that!" response. This tool of "thinking like the other" is offered to save time, create cohesion and end frustration. It should not create more frustration. Why not end every idea with a "She'll love this!" response? Personifying this new role can inspire and push you to thinking more creatively.

Director Revisions

It seems we have forgotten about the director during this revision phase. Are we implying that his work is done while the poor designers go off to dark corners and sketch? No.

This phase is where they rightly earn the title, director. They are literally directing all elements into a final product. They are guiding, funneling, and purifying design ideas to make them all mesh. They are steering revisions to flesh out the world of the play. They are directing all information to assist the others in their creative process.

He must be at his savviest at this stage of the process. As director, you must learn how each collaborator works outside of the production meeting. The dynamics established in the conference room are sometimes very different than the dynamics in private. You must balance all of the information you glean from your separate meetings with collaborators and interpret those decisions when moving on to the next.

This is also the phase where one's leadership muscles may have to be flexed in terms of maintaining control by putting out possible collaborative "fires." The director must be brave enough to make final decisions that may not be popular within the production team in order to serve the premise. This does not mean that you must prove that you are the leader, rather, it means you direct the conversations of collaboration to avoid wasted time and hurt feelings.

This revision phase is also a time for the director to reexamine and revise anything that still seems to be raising concerns amongst the designers. It is a chance to examine these elements collectively and, if necessary, rethink the direction of the production. Can a designer's idea now change the style of the piece in a better way? If so, revisions are essential.

We hope that a director will have enough forethought so that this seemingly whimsical approach to directing does not occur. Sometimes though, it is vital to join the designers at the figurative drawing board. Many times a designer imposes a change that must be respected as well. Whatever serves the production best is the ultimate judge.

Editing and Shaping the Style

In love, life, and design this homily holds true—sometimes it is better to leave things behind. Sadly, in any production it may mean your best idea!

You have presented an amazing concept...no one denies that. What they do deny, however, is that this concept works for this show. It is always tough to be a director or designer hearing the news that your great idea doesn't work.

Sometimes it is worse to see this amazing idea adapted then molded and finally handed to another. It is no longer your idea. It is just some element that has been destroyed because they didn't get it!

There are several ways to try and tailor your great idea to shoehorn it into this new world. You can offer portions of the design. You may shrink the scope of the idea. You can change the color to match the costumes. In each instance, you are creating a watered-down version of the concept that will not leave you in the least bit intoxicated when you see it on the stage.

If your collaborators have presented honest and realistic reasons why your proposal won't work using support from premise, budget, and audience; you must accept their decision. Remember though, you have this great concept ready to go when the right production does come along. All creators have a morgue of ideas ready to be revived at the right time. ❊

Final Approval

The revisions of the basic designs have occurred. The team agrees that they feel proud of their contributions and everything works cohesively.

The team does one final check.

The premise is supported subtly and fully.

The research is evident in relation to the truth of the design.

The revisions finally agree in style and tone.

Every element presented in the production is justified.

The audience will experience the production the team imagined.

There is harmony.

For a few hours at least.

A whole new wave of problems is about to hit the production team. Questions and miscommunications abound.

CHAPTER 6

COLLABORATION THROUGH REHEARSAL

This chapter will show you how to

- adapt your collaboration;
- defend your collaboration;
- allow others into your process;
- react proactively throughout rehearsals; and
- maintain your collaborative nature during technical rehearsals (techs).

Up until this point, the production team has been secluded while finalizing analysis, research, and design decisions. Their toil has resulted in a cohesive design of which all can be proud. The insular phase of the design is done.

As casting and rehearsals begin, a whole new slew of problems occur when real people begin to inhabit the imagined world created through design. Hopefully, the designers and directors have anticipated this occupation. But no amount of forethought can prepare the production team for these new collaborators, with new opinions and new demands. How can this carefully cultivated collaboration amongst the production team continue and thrive during this difficult time? How can designers collaborate through casting, fittings, and construction? How can a director prioritize actor and designer wishes? Especially during technical rehearsals, how can one remain calm, collected, and collaborative? We have a few suggestions...

Changes through Rehearsal

As the rehearsal process begins, the director's attention naturally veers from the designers toward working with the actors. The designers' attention naturally veers from the director toward the practical side of the design with construction and patterning filling their days. During these divergent agendas, it is imperative that directors and designers remain in contact and willing to continue in their collaborative efforts. Inevitably, things will change.

The expected changes should not be a mad scramble to rethink every decision, but rather a continuation of collaboration in order to accommodate the new members (actors and technicians) of your production. If a director casts a role in a completely different fashion than originally conceptualized, because an actor interpreted the role in a novel way, that decision should be honored. Similarly, if a designer has run into a snag over the logistical construction of an element in the design, the director must be willing to adapt. However, if a designer or director continues a constant pattern of such revisions, the production team may certainly ask those involved to stay with what has been approved.

There are two schools of thought on this usage of the word "approved" when it comes to collaboration. The first is that the design is finalized and will not alter. The second is that the design is always fluid and can shift until opening night. Both of these philosophies seem dangerous to us.

Altering final approved designs within reason should be allowed. What is reasonable? A director retooling the concept mid-rehearsal is unacceptable as is a designer who won't budge on the size of a platform. Such extremes must have compromises within them. A director may certainly rethink a key moment, but, depending on the budget or build schedule, a new concept may be impossible. As a designer, you may certainly stand firm to your vision, but you should be ready to accommodate reasonable requests that enhance the production.

> *You:* You are saying two things there! Who decides what is reasonable—the director or designer?
>
> *Us:* We were hoping you wouldn't ask that. This is a case-by-case decision. Fair judgment is needed. If the director has a reputation and penchant for revising the work of the designers, then ask the director to sign your drafting or renderings with the word "approved." If a designer has a reputation and penchant for being inflexible to any changes throughout rehearsals ask her to hold off on those elements that you are unsure of. Suggest that the designer draft or pattern those elements that you know will not alter. It is all a matter of bargaining.
>
> *You:* What if the collaboration breaks down because of such requests?
>
> *Us:* Simply state your case as designer or director in as calm a manner as possible. Actually consider this collaborator's request before dismissing it. If the revision will enhance the premise and make for a better production, then see if it is feasible. If the unwillingness to change the design is fair due to time and money that has already been expended, then respect that decision. We have experienced times in our collaborations when one of us has had a great idea close to opening night that

would really improve the quality of the show. Usually we look at each other and say "Yeah, that would have been great! Next show."

The sometimes emotional attachment to elements of direction or design is difficult to part with, but, in this phase, a new idea that better answers the needs of the production must win out. If the director loves the created moment occurring upstage, yet that moment will be diminished because of the limitations of the light plot, that moment must be adjusted or scrapped. The same is true for the designer who loves the gown that completely goes against the newly imagined moment at the end of the act; the costume designer must regroup. This sort of mercenary editing must be part of the collaboration.

Collaboration through Rehearsals

With designers and directors on divergent paths (that will hopefully meet up again) there is a sense of being removed from the production. As the designers begin to work with their shops, the director is removed from that process as a result of time intensive rehearsals. Designers may be drafting, preparing paint elevations, or involved with a new project. This disconnect can be salved by closer contact on both sides of the production.

One way to ensure continued collaboration is for the designer to maintain an active role throughout the casting process. Designers who witness auditions have a better understanding of what the director is looking for and what the actor will bring to the role. Discussion of casting choices with the director may inform and transform designs in all areas. What Actor "A" brings to the role of Mrs. Lovett is an elegance that the director did not originally see. Will the costume designer change slightly from the garish "nouveau riche" feel in Act Two? Will Mrs. Lovett's parlor not be as ratty as originally rendered? Is your Fleet Street a more fashionably lit part of town now?

This is also a time for the director to share discoveries with the designer. The director, unable to find the right actor in the given casting pool, casts against type and wants to play up a new look in the costuming. Suddenly you have an Ado Annie who CAN say "no," a Peter Pan who is afraid of heights, or a 6'5" Tevye, which means the roof must be raised. No matter what the issue, the designer must be made aware. Even if the director can't fathom HOW these decisions could affect the designer, include this new information in the rehearsal reports.

Clarity in Communication

Rehearsal reports are a daily way of connecting the collaborative team. Rehearsal reports should be succinct, clear, and thorough. Reports are usually prepared by the stage manager to create a bridge between rehearsals and production meetings. The purpose of these reports is to gather questions, raise concerns, or reiterate safety or practical issues throughout the process. For example, one entry of a report may simply state, "Actor 'B' will sit on arm of chair." This may not seem like much information until the designer realizes that Actor "B" is uncommonly tall and wide. The arm of the chair must be reinforced.

> R: I went into a recent production meeting rather angrily. I was wondering why my designers (including Kirk) were not responding to the MANY questions that filled the rehearsal reports from our production. I felt I couldn't go on blocking or using certain parts of the stage until my questions were answered. When I asked for the answers to all of these questions…
>
> K: I told him that we HAD been answering each question. It seems that the stage manager assumed that Rob was receiving these replies so he did not share the information with him. Rehearsal reports only work when everyone reads them.

Actors in the Design Process

In earlier chapters we discussed the importance of letting the actors in on your design process and creating the living museum in the rehearsal room. This sort of bridge between the design and acting world is necessary. So many times the relationship between these two factions can be thorny:

The actors say:
- "She is making me look fat because she hates me!"
- "He doesn't realize I can't sing while hanging from this rope in this corset!"
- "They can't even see me in my big moment!"

The designers say:
- "Her posture is ruining my costume!"
- "Another out of control actor broke my props!"
- "Why can't she find her light?"

In all cases a better understanding of the "enemies'" craft would solve misunderstandings. No costume designer deliberately sabotages someone's look; conversely, no actor deliberately sabotages a costume. These rants usually relate to something getting in the way of the imagined final product. Simple, clearer communication (usually through a director) is best.

Filtering requests through the director is important for several reasons. The director has the most knowledge of the actor's struggle through rehearsal to get comfortable onstage. One word about the sloppy posture can push back weeks of progress of the actor in educational theatre. Conversely, only the director knows how much work the costume designer spent finding and hand sewing the trim to give the costume "movement" that the actor HATES. These blurted judgments may incite an unwanted reaction from the cast or designer. Must the director mediate each debate? Certainly not. The director, however, must terminate the friction caused by unthinking comments during these asides.

For academic theatre another important element is to allow actors to become part of the design process by volunteering their time in the construction phase. Although it might seem impossible, actors who are made aware of the need for such help gain appreciation of aiding in the creation of the world of the play. It could surprise you. Ask actors to help and they just might do it.

A designer run is usually the first time that a director and actor have shown their work. It is imperative that a designer responds to actors as well as the director. Too much effort has been put into the process at this point to merely start into designer feedback. Always respect the process.

As a director, one should always volunteer free time in the shops despite any lack of skills. The opposite is also true: a designer who is a regular fixture at rehearsals can usually answer the simplest of questions or explain the function of the design. The trust and respect by breaking out of this "us versus them" mentality makes for a holistic collaboration.

Actors' Influence   — Safety   — ease of movement

One of the happiest days for the director is when the actor knows the character better than the director. The cliché line "My character wouldn't do that," is usually the best thing for a director to hear (within reason.) It is also a designer's worst nightmare. At this point in the process the actor is consumed with one element of the show—THEIR CHARACTER. The actor's job is to tirelessly dream about all elements of the character's life in order to fully inhabit this person. Ideas have begun to flow and the rehearsal reports are filled with requests for new props such as a cane, a moustache, and a tiger. A director can certainly edit some of the wildest whims, but should also support this active collaborator's discoveries that enhance the production.

How much is too much? When does it go from amazing discovery to selfish preening? Who decides what to support and what to discard? The director is the first line of defense between actor experimentation and design overhaul. Good ideas must be encouraged in rehearsal and the designer should be made aware of this new path. The final decision should arise from discussion between actor, director, and designer once they have seen the idea in action. They should entertain any and all options of implementation or exclusion. The questions regarding these new discoveries that affect the design include:

- Does this new element better illustrate the premise?
- Does this new element enhance the production?
- Does this new element reveal something relevant about the character?
- Is this historically accurate?
- Is this absolutely necessary?

If the answer to each of these questions is yes, then the actor's new discovery should become part of the design. Along the same line, actors who develop a "relationship" with a rehearsal prop or rehearsal shoes and want them for performance should not be mocked but rather celebrated for their ability to commit so fully to something representative. This sort of security for the actor is something that needs to be recognized and respected by the director and designer. As always, an actor who goes to extremes with such requests can understandably be labeled as difficult. Ask the above questions again and if the answers are affirmative, the request may be met favorably.

The dangers don't only lie in the actor's relationship with rehearsal furniture and costume pieces. Actors may be better informed if the director is able to communicate what the ultimate design will be. The designer should be present and willing to provide descriptions of particularly confusing elements of design as early in the process as possible. The director must also be able to describe in detail the function of the design to its fullest capacity. For example, the black bench will eventually be a red velvet pouf with unlimited amounts of blocking opportunities.

An actor who critiques the aesthetic of the design certainly oversteps the boundaries of acceptable behavior. An actor who rails against your railings or judges your jury box for reasons other than safety or ease of movement has no right to do so.

Other ways to marry designers and actors during this phase may include a scenic designer's walk-through. The scene designer could give a historical walk-through of the set and explain the architectural elements of the world outside of each and every entrance. A prop master could include the actor when decorating the character's apartment. Directors may ask actors in smaller roles to write biographies or answer questionnaires to assist the costume designer in fleshing out the very limited stage time for these characters. The insight gained in such exercises empowers the actors and helps clarify and deepen the design. Moments like those, which bring together the actors and designers, enhance the collaboration. This is what everyone should be aiming for.

The director must become a regular visitor to the shops. During these visits any embellishment or alteration of the original design can be seen and discussed with the previous questions also being asked. The director's presence should not be hawk-like checking on the progress of each phase of design. Rather, a director should experience first hand the difficulties, successes, and morale of these areas of the production. Even though a director may be technically challenged, time volunteered to base paint or sew buttons contributes to the production of the whole; not one created by separate camps.

Issues in Design Construction

Most challenges that occur during the rehearsal phase are known as "construction" issues.

Construction Issue #1: No matter how well the team has planned, issues occur when actors discover design challenges with choreography or blocking. For example, the dancer who suddenly needs a new skirt to perform the choreographed fan kick. Or, the rolling unit upstage, which now seems to overpower the tiny actor downstage.

Construction Issue #2: No matter how well the team has planned, issues occur when realizing the design. They may be budget or material issues. For example the cost of the chandelier in the rendering puts the production way over budget. Or, the material for the fabric trees is synthetic and can't be dyed to the proper color.

In both of these construction issues, bargaining takes place. If the skirt is already constructed, can a slit be added or can the kick be changed? If the rolling unit has been drafted, can the actor's placement be adjusted? If the imagined chandelier can't be found, can a suitable replacement be borrowed from another theatre or can a similar replacement be found within the budget? If the trees won't take the dye, can they be painted or can dye-able materials be found? There are often several acceptable options to most issues. If not, can the problem element be cut or rethought without reducing the integrity of the design?

The Lighting Designer in Rehearsals

We haven't made much mention of the lighting designer in this section of the book. Usually, this is because the lighting designer is spending time at home constructing the light plot. When in actuality, this is the time when the lighting designer must be most involved with rehearsals. It is the best time for a lighting designer to actively participate in order to create striking moments in the show. A lighting designer, present at rehearsals, asking questions of the director makes a wonderful practical addition to the process.

At the moment the lighting designer sees the mood and style of each scene, a plot to showcase important moments of the production can be created. Moments not addressed in the production meeting but discovered in rehearsal may be highlighted. It is much easier to shift downstage two feet to the hot spot in early rehearsals rather than adjusting during the stress of technical rehearsals.

Another way for directors to involve the lighting designer may be sharing an overview of the blocking patterns to the designer. Several illustrated diagrams of each scene are a great help. Drawing the show's traffic patterns on several copies of the ground plan ensures that a lighting designer is ahead of the game before observing any rehearsal.

Production Meetings through Rehearsals

Even with active collaborators attending rehearsals and responding to rehearsal reports, in academic settings and beyond, it is necessary to have regular production meetings to address questions that have arisen from each member of the production team. The agenda usually remains the same for each meeting, but the tenor of these meetings has turned from the creative to the practical.

Always begin a production meeting with an update on the progress of the rehearsal. Too often this part of the production is never addressed in a production meeting. Most times these meetings solely concentrate on the design/technical elements. The meeting should begin with a frank discussion of how the rehearsal process is going. A director should talk about discoveries or challenges with certain moments in rehearsal thus reminding the production team of the expanded community that has been created in the rehearsal room.

This sort of inclusion on the process makes for an interesting springboard for the rest of the updates from the design areas. These meetings should address unique issues that inform all members of the collaborative team. While some may balk that the design is complete and these meetings are unnecessarily redundant, the minutiae of these meetings should (hopefully) pave the way for a smoother technical rehearsal process.

In the final production meeting before technical rehearsals, prepare a plan to accommodate the needs and schedules for all collaborators to have a fair amount of time to present and hone their element of the collaboration efficiently and effectively. Expectations for each technical rehearsal should also be set at this meeting. "On Tuesday of tech week, we plan to have_____." This sort of spoken promise should be honored while also allowing wiggle room if one of the design team needs more or less time than expected.

Effective Technical Rehearsals

Now that the rehearsals are nearly through, the scenery is loading in, the costumes are getting finishing touches, and the lighting is being focused—the technical rehearsals are imminent. This stressful time is unlike any other part of the collaborative process. The designers are now the prime motivators of the collaboration, the director merely a respondent. The sketched elements of the designers' imaginations finally blend into a tangible cohesive whole.

The technical rehearsal is purely that—a rehearsal for the TECHNICAL elements of the production. It is not time to be spent working with the actors. While acting notes are common in these rehearsals, a director should spend the majority of the time responding to the design elements and the designers. Even if you think of yourself as an "actor's director" your attention should be firmly rooted in the technical aspects of the production. It is the time to see if these individual elements work together.

> R: I hate technical rehearsals. No matter who I am working with, I always
>   bring this energy of "This is going to be hell" into the theatre. It is not
>   from a lack of trust, but rather the countless techs (as an actor) that I
>   plodded through. And, WRONGLY, I always prepare my actors that
>   the upcoming event will be unbearable. It is a self-fulfilling prophecy. I
>   usually end up at opening night going, "Hey, that tech was easy." But the
>   ingrained distaste for techs make me a bad collaborator.
>
> K: My anxiety lies within the first two days. I have my list where I begin
>   to cross off the major paint notes and the functioning of the turntable
>   and the swiftness of the fly operators. I long for dress rehearsal when
>   costumes hit the stage and all of the attention is drawn away from me so
>   that I may concentrate on the "picky" and minute details. The stuff only
>   my mother would notice.

No matter what your feeling for techs may be, the smooth running of these rehearsals must be handled by the stage manager. In academic theatre, a stage manager may sometimes be a neophyte who lacks the requisite strength, respect from peers, or compassion to wrangle the many forces of the technical rehearsal. As a director, perhaps introduce certain elements to the stage manager so they can practice before techs begin. Have the new stage manager run the sound cues or call light cues during rehearsal. This sort of preparation for a novice ensures smoother techs.

In this newly configured chain of command during techs, the stage manager takes the lead and juggles all elements of the production...including the director. This new hierarchy may be confusing and frustrating (for the director especially), but the stage manager needs to be empowered to call a stop to the run-through by yelling, "Hold, please!" The stage manager is now the lead of the production team. All concerns, questions, and comments must funnel through him.

Setting Agendas

At our university, we handle techs in roughly the same pattern for each show. One technical rehearsal is devoted to a single design element. It works for us, it does not have to for you, but the idea of using each tech to introduce a new element seems to give ample time to repair or rethink. On our first night in the space, the major elements of the set are in place. That first night is a spacing rehearsal that allows the director and actors to see if the imagined blocking will work in the space. The set designer is usually there to introduce the set and its workings. This night is about safety and any practical issues involving the set that impede the movement of the production.

The next rehearsal, the lights are introduced to adjust blocking, synchronize timing for cues, and examine the look and feel of each scene. The next element and rehearsal is devoted to the costumes. The same agenda of keeping the production moving forward including quick changes is the goal of that rehearsal. The next rehearsal introduces the makeup and hair for the actors. The last rehearsals are devoted to the cohesion of the collaboration.

Throughout each of these rehearsals, the designers and the director are fine-tuning and adjusting their work to make the production successful—unified and supporting the premise. During the onslaught of design elements, the director should focus attention on the introduced element and how it meshes with the other elements of design. Of course, everyone on the production team hopes for success, but we have never been involved in a production where notes are unnecessary.

A Note to the Director from Rob

The director sets the tone for these rehearsals. Stress is easily noticed...as is fear. Be sure to project the correct tone for your production. Always, get to know your crew and thank them for their work. Always make sure to give positive notes as well as negative to your designers. In this sort of hurried introduction to one design element it is easy to dwell on all that has disappointed you.

A Note to the Designer from Kirk

Remember that the director is dealing with all elements. Make sure that your element is in the best shape it can possibly be. Help to incorporate the director's needs as you guide the crew from one shift to the next. Make sure to give positive notes as well as negative to everyone involved. Most importantly: PRIORITIZE. I cannot stress this enough. Spend the time during tech notes to prioritize your needs with those of the entire production team. Deal with what is essential to the production. This may also be the time to lend a hand to other collaborators who may need your help.

Director's Notes during Technical Rehearsals

So, the technical rehearsals are underway. Notes are about to be given and feelings are bound to be hurt. How to best handle this?

We both agree that, in a technical rehearsal, the designers and crew should ALWAYS get their notes before the actors. In this reduced rehearsal process for the designers and crew, they must be addressed first since they usually have the most adjustments and the least amount of time. The actors have been in rehearsals for weeks; let them change out of costume while they wait for you to finish your discussion with the designers. In all cases, be reasonable. If the discussion with your designer following notes is going to be a LENGTHY one, give your actors their notes and then return to the discussion.

Designer notes should be a give and take of what worked and what needs work. In this phase, a radical rebuild of one element may take numerous hours and dollars. Always seek to modify rather than scrap. Examine the element and see HOW to make it work rather than how to discard it. In seeing (and hating) the "Hot Box" dresses for *Guys and Dolls* in action for the first time, ask a few simple questions (either as designer or director):

- Is this a fair representation of what we talked about in the production meeting?
- Did this element turn out the way I had hoped?
- Would I be proud to include this in our production?
- Is this element capable of being salvaged through alteration?
- Once altered, will this element be acceptable in the limited time we have?

If all of the answers to these questions are "No!" you had better head back to the fabric store. If more than one of the answers to these questions is "Yes!" you need to begin discussing immediately how to reconfigure.

Maintaining Respect

- How best to break the news that you dislike a designer's contribution?
- How best to break the news that you cannot deliver on the promised design element?
- How best to tell a designer you misunderstood when you last spoke in the production meeting?
- How best to tell the director you disagree with his picayune notes?

As in all exchanges imagined by the authors of this book, handle all collaboration with understanding, grace, and tact. Have understanding for the work of your collaborators. Respond with grace for the frayed nerves of a director who spent weeks in the rehearsal room slaving to make the actors inhabit the text. Have tact with the designer who spent hours trying to ensure that this production looks beautiful. COLLABORATE with all involved who are there for the same purpose: to create the best production possible.

Very rarely does anyone's laziness hurt your final product. Why would one be lazy in theatre? Is it the regular hours? The overwhelming glory? The oodles of money? It is pretty much guaranteed that perceived laziness is really disaffected exhaustion. The variables of your exhausted technical director, "refusing" to make the exit downstage left work, may well be the fifteenth attempt to solve this problem before you became aware of it.

If, of course, this sort of empathetic approach to your collaborator's contributions is not working, there is always the tough approach. Try pushing to get each issue solved right up until opening. Account for each perceived problem and define the solution. This is not the place to bring up mismanagement, but rather a time to problem-solve and push forward. Blame and anger can happen after opening night.

If anger or stress becomes too much a part of the technical rehearsal, reconvene the next morning when fresh heads prevail. The great equalizer in each of these debates is the premise: Does the argument serve the message of the production. Is the debated element the MAIN thing missing from the production? If the stress over unfinished details seems to overwhelm the tech note process—prioritize. Spend your time at home that night creating alternatives or means to solve the challenges.

If the notes end as hoped, create a list of priorities along with a schedule for the following work day and tech rehearsal. Begin with the elements that affect the actors or impact other design elements. Save the picky details for the day before opening. The baseboard can easily be added a few minutes before house opens. Just make sure that this addition is on the list and will be completed. The director should be involved in this prioritizing, but can also be overruled by a designers' understanding of workforce and materials.

Opening Night

Opening night is upon us. The production has been running smoothly. The stressful moments of the technical rehearsal have been diverted through a respectful understanding of the hard work of each collaborator. The team all breathes a sigh of relief and swells with pride as the curtain is about to rise. This is the time to recognize and appreciate the work of everyone involved.

How best to show your appreciation?

The best opening-night gift may be the promise and hope of future collaborations. But before you swear to work together in the future, take a second (or two) to let the production open. Let your experience sink in. Take a moment and breathe as we head into the final phase of collaboration.

You mean there is MORE? Yes. A final chance to review, recap, and rethink your collaborative process.

CHAPTER 7

THE COLLABORATION, POST MORTEM

The production has opened. The response was what you hoped for. You see the production team's final product and you've marveled that this project, begun many weeks ago, has been successfully handed over to its intended audience. You have a few seconds to breathe before the whole process begins anew with a different production team.

That downtime between productions is the perfect occasion to examine your recent process. This respite offers opportunity to analyze the best and worst moments of the production and your responsibility in its ultimate success or failure. The post mortem evaluation provides the perfect opportunity to scrutinize the collaborative process and set goals for upcoming productions.

This chapter will show you how to

- stay collaborative following opening night;
- analyze the successes and failures of the collaboration;
- ensure future collaborations; and
- ensure healthier collaborations on the next production.

Defining Success

More than once in performance, we have observed a beautifully designed production that did not match the level of the student actors. More than once we have strained to see the faces of actors in pivotal moments. More than once we have seen constricting (yet historically accurate) costumes hinder a show. More than once, we have seen a director value stage composition over truth. Your work as designer or director while aesthetically pleasing may not serve its main purpose—to immerse the audience in the production.

The first arbiter of a production's success is the audience's reaction. We are not implying that if an audience laughs and applauds at all of the right places that your collaboration was successful. Rather, we refer to the entire experience of independent visitors to the world created by the production team. Was the reaction in their visit what you intended through appropriate, cohesive, and supported decisions? We imply a production where the elements agree. We imply that the style of acting and cut of the tailcoat correspond. The passionate mood of the scene is reflected in the lights. The doors of the farce slammed at the appropriate moments. The director staged the climax as such. Most important, that the premise of this unique production of the play was served by the production team and was received as intended by the audience.

Short of stopping strangers and asking them about your element of the collaboration, what can one do? We suggest a true viewing of the performance later in the run of the show if possible. We ask you to strip your mind of the bias and personal attachment to the production and simply experience it as a true audience member might. Of course, this is what directors and designers should do during rehearsals and techs but rarely happens. Some designers and directors are so sick of seeing the show during techs that they seldom see their production in performance. This mitigates the essence of collaboration.

We suggest devoting an evening at the theatre with a clear agenda: to simply experience the production. Remove all memories of budget, creative and technical woes, and simply watch the show. Upon entering the theatre give yourself over to the experience and define its successes and failures rationally.

If you, like many artists, are unable to divorce yourself from your creation, choose a total stranger to you and the production to study throughout the show. Choose Person "X," seated in the seat in front of you, and try to view the production through her eyes. When her attention wanders, define its root. When she turns to her friend with a knowing glance, trace the cause. If she laughs at a particular moment, inspect the source. What was your contribution to these reactions? A funny script only works when elements agree. A moving moment only blooms in a truthful environment. A wonderful script may be damaged through inflated design elements. A play that made you laugh out loud when reading it may land with a thud in performance because it was overly directed. Be hyper-aware to her reactions.

Only through thoughtful analysis during a dispassionate viewing can a collaborator truly define the success of the production. How a show reaches an audience is the true test of collaboration.

Reviewing the Review

One audience member you can always count on to describe the experience of watching your production is the reviewer. The dreaded review is an accurate and valuable appraisal of the production. The fact that it has been published is both blessing and curse.

We do not imply that every review is sacred and the definitive word on a production's success. We do imply, however, that a review is one person's reaction to the work onstage. Whether you agree or not, or whether you think this critic has an agenda against you, the theatre, or the world, the summation of the experience is true . . . for this particular person.

You should read reviews and gather audience reaction carefully and with as little ego or judgment as possible. See what you can gain from anyone's response. Although you don't always have to concur, you should examine.

> *R*: A reviewer once remarked that my blocking in a certain production had the actors' backs to the audience too often. I dismissed this review saying this show was produced in a thrust theatre and that sort of blocking will always be a given. I had spent rehearsals bouncing from seat to seat to make sure the show was nicely sculpted from each area of the theatre. I prided myself on my equitable distribution of "moments" to each area of the theatre. My defenses were engaged until I realized, her opinion of this element was strong enough to write about and I should simply accept her critique as what it was—a recap of her experience at the show. Her enjoyment of the production was marred by the fact that at key moments she felt excluded from the story. I went back to the show and sat in the same seat as she and understood exactly what she meant. I did not agree, but certainly understood her experience.
>
> *K*: I received positive reviews for operettas I have designed in the past that referenced the ambiance and "flawless" environment of the design. As I look at those reviews, I have to question whether I served the script. While the attention was flattering, that was not a desired review. I would prefer not being mentioned in a review of the production where all of the elements worked cohesively.

We needy artists thrive on these written evaluations of our work. The old adage that one shouldn't read reviews is poppycock. You should read reviews but with a careful mantra of "it is only one person's opinion" running through your mind.

At our university we have students write responses to each production in our classes. These "mini-reviews" are an excellent way of evaluating one's work. While majors are busily running the show or building the costumes, we look to the non-majors for the more truthful response. Did the accounting major seeing his first live show understand the production team's intent? Too often in educational theatre we forget that the typical audience member does not have the research base or breadth of theatrical conventions that we do. A production should not come with a glossary or a five-page director's note, it should be judged on its ability to connect and affect said accounting major.

Peer Review

The best way to get feedback is by finding people whose critiques you trust and respect and ask them for an honest response of what worked and what didn't. These touchstones to the truth can be theatrical peers, appreciators of the form, or simply your nextdoor neighbor who tells it like it is. This truthful discussion can only come after sparring in debates about the merits of past-shared theatrical experiences.

Personal Review

By collecting data from unspoiled viewing of the production or through someone else's eyes (audience member, critic, or colleague) a clear picture of the successes and failings of the production should emerge. Now is the time to review the collaboration on your own.

One way to do this is to review your research, script analysis, and implementation into the final product. Did you achieve your goal of involving the audience by placing the main character's chair downstage? Did you assist in defining the climax of the show by raising the intensity of the lighting cue? Did your gown achieve the freedom of movement explicit in your research images? Did your work support the premise?

Review all production meeting notes and rehearsal reports. Did you fulfill your agenda from the first meeting? Did you respond to the repeated questions in the rehearsal report? Are you happy with the decision to cut the problematic element? Was it worth standing your ground for another such element?

Your answers to these very important questions may assist you in gauging the success of your collaboration. Only through thoughtful analysis of the final product may one truly complete the collaborative cycle.

Individual Review

That cycle continues into an examination of your work with each member of the team and everyone connected to the production.

Will your fellow team members work with you again? Have they or will they refer you for a future production? Would you choose to work with them again? If not, why did your styles clash? What is your responsibility for this problematic relationship?

These same questions must also be posed on an even larger scale. You must examine your relationship with EVERY person connected to the production. Did you lead your crew effectively? Would they work for you again? Do you know any of the actors' names? Would you choose to work with them again? Did you thank people for their time and patience when you were pushing to perfect your element of the collaboration? What was your responsibility in having to push?

That sort of review process should occur on some level following each production. If patterns appear in your reviews, address them. If you never once spoke to the actors in techs, change that. If you have to corral new crew for each project, adjust your style in the shop. If you never get asked back to direct, amend your schmoozing ability. If you never get a referral for another job, alter your style.

Although referrals and securing future work is paramount, one element of the collaboration must be addressed—your reputation. Your reputation precedes you and, in most cases, is never made known to you. It is whispered or e-mailed but rarely shared with you.

Your reputation is usually created in moments of highest stress. Review those moments of the production that were most stressful and examine your behavior. Did you "lose it" or did you rationally deal with the problem? Your reputation can usually be traced to those defining moments. Why not change your reputation from someone who "freaks out easily" to someone who is "graceful under pressure?" Reputations can be changed. History cannot.

R: I think my reputation is that I am demanding yet fair. I can live with that.
K: I think my reputation is that I am always prepared and you get what you are promised. I can live UP to that.

Team Review

One final meeting may be necessary in order to close the chapter on the production. This meeting, known as the post mortem, is a final chance for all to reflect on the collaboration. The term post mortem conjures images of forensic scientists examining a cadaver for clues—and that (in a less grisly way) is exactly what this event should be. It is a final examination of the corpse (the production that has closed) for clues regarding successful moments of collaboration.

The post mortem comes in many variations. In some places everyone involved in the production meets to discuss the issues of the show. Two outcomes usually occur when a huge group has gathered for such a talk. The first is a lovefest of compliments and kisses. The second is a blamefest of complaints and finger-pointing. Neither is terribly productive. While this large meeting can certainly yield some interesting feedback, they are usually too unwieldy and understandably someone feels cheated because of the group dynamic. These events are an excellent way for producers to gauge the overall response of those involved with the show, but frequently the core team is on the outside of such a talk.

The second is a gathering of the key players of the production team. Be sure to include shop supervisors and assistant stage managers who can offer feedback from crew and actors. The post mortem agenda can be as free form as just letting the conversation flow from one subject to the next or as rigid as reviewing the production either chronologically, departmentally, or collaboratively. This is also a chance to discuss communication skills individually and as a team.

This meeting should not be a waste of time, if you, as team member, have a purpose to truly discover what others felt about your part of the collaboration. You have already gleaned personal and audience reaction, now it is your chance to let your co-workers discuss your performance. Don't defend; rather, remain open and active. Remember that much more can be learned by simply listening to their opinions.

When it is your turn to discuss a fellow collaborator's performance, remain honest and cognizant of the issues that molded this person's accomplishments in the team. The post mortem can be a dangerous balancing act if there is someone with whom you did not enjoy working who is seeking honest feedback. As we suggest throughout this book, truth tempered with civility is always the way to go. Your reputation is secured in diplomatic moments such as these.

If your organization does not participate in post mortems and little support for such an event is forthcoming, schedule individual meetings with the producer, production team, or shop supervisors to gain personal insight to ensure growth. The agenda is purely to discover your successes and failures in your co-workers' opinions or to review the communication between the team. Again, ignore the impulse to bash or blame others and use the time to focus on ways to improve your next collaboration.

Personal Goals

The collaboration is almost complete. One final item to tend to—personal goals for the next production. It seems evident that after such extensive evaluation that you should know what to do differently next time. Not always true. Take the time to itemize clear, prescribed, and achievable goals. They can be as simple as "do not lose temper" to the radical "rethink my process." Review the major tenets of part I of this book to create bullet points for such a list.

- Did I distance myself from my collaborators through ill-chosen vocabulary?
- Did I analyze the script specific to the production?
- Did I research our production fully?
- Did I address the director's/designer's needs during revisions?
- Did I effectively communicate my ideas to others?
- Did I remain collaborative during rehearsals?

This list does not address all elements of collaboration, by far. Concepts stressed throughout this book like respect and trust between the collaborative team should be included on any list. The details and examples included in each chapter must be revisited as well. All of these elements in combination with a true openness to collaborate should create a clear path to success.

R: When I am preparing for a new collaboration with a colleague I worked with in the past, I like to review production photos. My visceral yet removed reaction to these photos reminds me of things I want to stress or avoid in our new collaboration.

K: When I am preparing for a new collaboration with a colleague I worked with in the past, I like to review my notes. I want to be prepared to respond to the personal quirks, questions, and quips of the specific collaborator to expedite the design process.

Healthy Collaboration

Through careful examination following each collaboration, one can become a better member of future production teams. Those who explore the defining moments of successes and failures of past productions will find themselves growing as artists and can begin cultivating future collaborations.

Healthy collaboration has one major goal, to create a unique production. Healthy collaboration also has one fringe benefit, securing future work. Positive and active collaboration continues in the weeks and months between gigs as you seek new work. Knowing that your reputation as collaborative artist is assured then future work should be forthcoming.

Each time the phone rings with a possible new project the same hope/dread infects us all:

Can I achieve that feeling described in the introduction of this book? Will this project and the production team be willing to collaborate? How can I get to that dizzying exchange of ideas and information to make collaboration on a cocktail napkin truly possible?

PART II

COLLABORATION IN PRACTICE

R: With this production of *The Life*, we were very excited to put our theories into practice. Our goal was to create a worthwhile and positive experience. We were excited to step up and see if we could make our theories work.

K: Yes, and I think it was really frightening, because we both complain—a lot—about non-utopian collaboration.

R: Our goal throughout was to push each other and to be respectful. By actually talking about this book we learned quite a bit about the best language to use and how to treat another person. But we also failed occasionally. And I think that's the point of the book...that if you give yourself a solid foundation, you can fall back on it after you falter for whatever reason.

K: That's comforting.

R: Don't you think entering a collaboration exhausted is a real factor in your prep work for this show? It is also a probable factor at some point for everyone reading? I mean, it happens.

K: Entering any situation exhausted is a problem. So, you have two choices—sleep now or push through it. You're going to have to be a much more active listener up front while you're playing catch up in order to support the director's vision while exploring your design process. It's a part of ANY collaboration. It's life.

R: *The Life*.

In the first part of this book we spent a lot of time helping you prepare for a healthy collaboration. In this part, we will explore the creation of an entire production in which we practiced the theories from earlier chapters. We hope to recount the practical application of what was explored in part I and honestly elaborate on the failures and successes of what happens when we practice what we preach. It is your job to discover the moments of desertion, distance, or double-talk when the authors divorce themselves from the theory espoused in part I.

The successes chronicled in this second portion prove how organization and clear vision can streamline the creation of a successful show. The vocabulary, timeframe, organizational skills, attitude, and clarity of premise combine to create a world in which designer, director, and actor want to partake. It describes that rare instance when producing a show hardly seems like work because of the investment of each collaborator.

The occasions when we eschewed elements of the first part of the book are also prime for examination in the following part. Budgetary issues, exhaustion, and time constraints tested the collaboration. Moments of frustration and cutting corners led to poor decisions between designers and director. All will be revealed.

As colleagues, we had some misgivings about working together for the first time. We had seen and respected each other's work, but collaboration is the true test of the artist. In that process every flaw and shortcoming is revealed. On the positive side, every spark of genius and compassionate connection can be observed.

In entering this partnership, we each had our own theatrical baggage. In past collaborations, we've either been disappointed by someone we knew, lost a friendship due to infighting, or decided never to work with someone again because of witnessing our collaborator at his or her worst. All of those fears were in the backs of our minds.

One caveat: we knew going into this production that we would be dissecting the collaboration. Perhaps our adhering to the utopian suggestions of the first part may have skewed our data. That may have played a factor early on, but was soon forgotten as the work began. For the most part, our collaborative styles did not change. Our fellow designers were unaware that they were part of this experiment. Their part of the collaboration was completely unbiased.

It was now time for us to test the ideas we had generated.

And so it began...

CHAPTER 1

PREPARING FOR *THE LIFE*

In this chapter we will examine

- choosing a show;
- creating the premise;
- overcoming challenges;
- analyzing the script;
- assessing the collaborators;
- exploring personal history;
- expanding the vision; and
- communicating the premise.

Selecting the Show

Collaboration for us began in selecting the production season for our department. We knew we would be working together on a musical, but which one? Some shows for the season had already been selected, so we had to choose a show that struck a balance between those titles. The shows had been chosen to correspond with MFA designers, other director's wishes, and the department's production history.

Choosing shows for our audiences and available actors is always a factor in the decision process. Each show in our season is a part of that whole and equally part of the greater theatre community in the area. Technical production capabilities and manpower play a major part in most show decisions here and elsewhere. Budget constraints related to scenery and costume construction fluster the best of collaborators.

Each phase of theatre is really a collaboration, and title selection is no different. Finding a way to work within the strictures and structures can be both a director's and designer's best asset. Knowing the limitations and possibilities of your production organization begins immediately. Our feeling is that any production can be mounted. The only question is—HOW? Your job as designer or director is to mount the best production possible within the limits set.

In recognizing manpower issues and all of the above, we also had a few other elements to inform our choice. As we sat down to decide the season, here were the givens:

- We needed to do a musical with very little dancing because our Repertory Dance Concert was rehearsing at the same time and the best dancers would be otherwise engaged.
- We needed a musical with modern dress because of the unique costuming demands of other period pieces in our season.
- We needed a complicated lighting show for one of our excellent MFA lighting designers.
- We wanted a show with good women's roles because (as always) the other shows in our season were men-heavy.
- We wanted to stay away from Rodgers and Hammerstein–y kinds of shows, having recently produced *Oklahoma!* and *Carousel.*
- Selfishly, Rob wanted to direct a musical that was edgy and challenging for Michigan audiences.
- Selfishly, Kirk and Rob wanted to work together as director and designer.

All of those factors played into our decision. Yet, there was one unnamed factor that was a major monkey wrench and saving grace when choosing the musical. This production was to be a COLLABORATION between the Department of Theatre and the College of Music. All complications implied in the joining of two producing units were givens here: The difference in vocabulary ("maestro" versus "conductor"), rehearsal times ("I need to save my voice" versus "You need to explore this moment with more rehearsal"), and logistical finger-pointing ("That is *your* part of the production!") were to be part of this undertaking. The monkey wrench was that the College of Music is understandably interested in productions that showcase the voices of their students while the Department of Theatre wants to showcase their actors. The positives of the arrangement were that this collaboration offered a larger budget than our normal theatre productions and both sets of our respective students could learn a whole new way of working.

Sleepless nights of listening to cast recordings, reading librettos, scouring catalogues, and researching on the Internet ensued. Happily, the point person for the College of Music is just as much a musical theatre buff as we are. So, the lists of possibilities were quite interesting. Titles such as *Sweet Charity* (too much dance), *On the Twentieth Century* (the operatic score would sway casting to Music students), *City of Angels* (too expensive to do the way we wanted to do it), *King of Hearts,* and *Bajour* (well, we like them) were bandied about. As you can see the people compiling the lists had their own collaborative agenda. We each brought our own history and taste to the table when discussing the show choice. This process was respectful because we had such joy in imagining the possibilities. Also, by spending so much time discussing the kind of show that we wanted to produce, the ultimate production would surely fulfill that.

As a group, we kept returning to the work of Cy Coleman. A discussion of his amazingly divergent career was what led us to *The Life*. When we hit on that title we all stopped. We found the perfect answer to our edgy, modern-dress, less dancing, women's roles, lighting intensive, un-*Oklahoma!* show!

Selecting for a Specific Audience

The Life is an adult musical by musical theatre legend Cy Coleman that depicts the seedy life of hookers and pimps in 1980s Times Square. Although most of the characters are trapped in "the life," the musical is an inspiring exploration of how friendship can offer an escape from a desperate situation. The musical won the Drama Desk and Outer Critics Circle Award for Best Musical and two Tony Awards in the 1997 Broadway theatre season.

The show is narrated by a street hustler, Jojo, who introduces the audience to a Times Square of the early 1980s, described in the show as a time "before Mickey Mouse came in and cleaned up the place." As the show progresses, we are introduced to a slew of prostitutes, pimps, and drug dealers—among them Queen and Sonja, two members of the "oldest profession." Queen's boyfriend, Fleetwood, becomes interested in the new girl in town, Mary. Queen decides to take this opportunity to escape from the life. Getting out of the life is made more difficult under the watchful eye of one of NYC's biggest pimps, Memphis. Queen spends the show trying to escape the ties to her coke-addicted boyfriend and her obligations to Memphis. With Sonja's help, she ultimately does so.

Once we found the show we wanted to do, we remembered that we were on a Big Ten University campus. Were we going to ask a bunch of eighteen- to twenty-one-year-olds to prostitute themselves, simulate drug use, and dress provocatively? Absolutely! As a group, we were behind the show, so our collaboration was forged from the beginning. As we defended the show—to University censors, theatre purists who complained of misogyny, and concerned music teachers who felt the score of gospel, pop, and Broadway standards would ruin the voices they had trained—we grew stronger in our conviction to be the first educational institution to produce this show. In answer to the critics, we suggested that the show was tame in comparison to other material students were exposed to, that we found strength in the women characters' wish to escape the confines of prostitution, and that we were excited to challenge the voices of classically trained singers, which would ultimately make them more flexible and marketable. Eventually we won over the detractors. From title selection until opening night there were approximately seven months.

Director's Preparation

[Since the rest of the chapter is about the director's pre-collaboration work, Rob will be writing alone here]

Immediately, I began the arduous journey of script analysis to define the production in the midst of several other projects. I tried to adhere as strictly as possible to the script analysis and research theories described earlier in this book.

The script analysis phase is where the director is most lonely. Although I love to explore scripts, I always wish I had people to bounce ideas off. In a perfect world, this discussion could take place with the designers, but I have found that most times the designers are tied up working on other projects.

Immersing myself in the libretto and score for *The Life,* I made many discoveries. I saw the nearly operatic structure of the show with its repeated motifs and recitatives. I heard the eclectic nature of the score. I saw the anguish of each character trying to escape his or her current situation. I saw the greedy ambitions of the antagonists reflected in action and in songs such as "Use What You Got," "Mr. Greed," and "People Magazine."

I also perceived traps in the script that I would have to address. I felt that the underdeveloped friendship of the two lead women rendered the song "My Friend" unearned at the end. I questioned the repetitious nature of abuse and reconciliation in the relationship between the lead couple. I hated the long running time of the show. I squirmed at the scene change music that broke the momentum of the piece. Again, all of these "flaws" are filtered through my sensibilities. Another director may disagree with all I have written.

Finding the Premise

At this point I was bursting with ideas. I had to assemble them into a streamlined message to guide myself to a dramatic statement to assist the designers and to focus this huge undertaking. I started with the premise.

I examined the title and dramatic structure (especially the climax). This is my shorthand to find the premise. In this episodic script with many major characters, I had to focus on the main character's story. It was the story of Queen, the reluctant prostitute. She begins the show freshly released from jail and returns home to find her boyfriend has spent most of the money they had saved. That money was to start a new life together. I followed her story to answer some questions. What does she learn? What is she seeking during the entire show (her super-objective)? I assembled all of these clues.

Then, I began brainstorming. I wrote down dozens of words that resonated with my experience reading the script. Other directors may have chosen completely different words or themes. My script was littered with underlined lyrics and random thoughts. My duty to the designers—to successfully communicate a unfied message—was paramount in this process.

So I knew the show was about escaping. The question became "What sort of escape?" It's obviously about money. It's obviously about using other people and selfishness. I went to the climax of the show and I saw the character Fleetwood uncharacteristically sacrifice his life for his girlfriend so she could escape. I also saw Sonja sacrifice her future so that Queen could escape. Those moments of atonement and sacrifice were in direct conflict with the narrator Jojo's mantra of using others to get ahead. I realized the show was very much about sacrifice, or, on the darker side of human nature, the lack of sacrifice. After arriving at a concept for the premise, I began the process of condensing it into a single sentence. After much deliberation and several false endings, I came up with my premise: Escaping from the life leads to sacrifice.

> R: Here is a really embarrassing moment for me as director. I was directing *The Uneasy Chair* by Evan Smith. At the start of the rehearsals, I was exhausted, having just come off a huge production. I was really unprepared to begin the show. I cursorily did my usual dramatic structure and premise work for my designers and actors. At our first rehearsal I sat through the read-through and really listened to the show for the first time and realized I had completely misinterpreted the script. COMPLETELY! I had misread the inciting incident. I had shoehorned a premise into the script out of hubris ("I can do this stuff in my sleep") and exhaustion ("Do I have to read it again?"). I sat there aghast that all of my table work was moot. I had a very capable cast who never mentioned the disparity between the written work I had given them and what I was saying. The entire process for me was catch up. Don't do that.

I was positive I was on the right track when assessing *The Life*. Now I needed to assess my problems and strengths as a director and the limited knowledge of my collaborators.

Assessments

- I would need to bring my skill of helping actors create unique characters to really flesh out the chorus roles.
- I needed to challenge myself to keep the show moving.
- I couldn't rely on my tricks of humor to deal with the raw emotion of the show.
- I had to make sure that I could speak the language of the College of Music collaborators (Again, adapting to another's vocabulary).
- I had to treat this gritty and salacious subject with respect.
- I had to make my actors and designers safe enough to dare to explore this uncharted area.
- I had to mentor a student designer to make sure he could feel entirely equal in the collaboration.
- I had to maintain the professional respect for my colleagues.
- I had to challenge myself and the designers to create something unlike the productions I had seen at our school.

Personal History and Collaboration

Luckily and damnably I had too much history with the show and its locale. I saw (and loved) the original Broadway production, but had little memory of specific design choices.

I also had lived in New York City, pre– and post–Mayor Giuliani. I remembered what it was to walk down that stretch of Times Square. I remembered that I was titillated by the neon and shadows in this crazily busy area when I was by myself, watching drug sales occur and tourists being accosted. I also remember being embarrassed by the obvious seediness and exposed sadness in the daylight when trudging with my family through this area. When Times Square was cleaned up a decade after the time of *The Life*, I mourned the loss of this area of town, while at the same time I was relieved that the "crossroads of the world" was safe again.

These issues became problematic in my discussions with my designers. I would refer to that costume from the original production or a specific building that I knew so well, but they had not experienced those things. I almost wish I did not have that background for the purposes of this book, but each director has a relationship to a show and mine was that.

I had analyzed the script, assessed myself and the production team while exploring my history with the show. Through this process, a director can't help but make some decisions that affect the collaborators. A director can become fixated on a passage in the script, see colors for the production, have a sensual reaction to the atmosphere, or create a defining metaphor for the production. It is only natural that a director concocts images from those hours spent alone with the script. The challenge is how to avoid designing the show entirely in one's head during the process.

Making Decisions Pre-collaboration

As I ended the script analysis phase, I made two decisions on which I was immovable (at that time). Not really a great way to collaborate. However, those moments helped me to define the production.

I knew that I wanted my characters and audience to feel trapped. This was of course related to the premise. I wanted them to be unable to escape. Not too general, eh? I knew I wanted the show to be assaultive and dangerous. This led me to my first major decision—the show would spill off the stage and into the house. The narrator addresses the audience at the top of the show; the textual support was there. By removing the safety of the proscenium, the audience is forced to empathize with or recoil from these characters and hopefully feel the trapped quality of the premise. This seedy world would envelop the audience in the ways I remembered from my time living in 1980s Times Square. This decision obviously would affect every designer. The scenery must be expanded, the costumes must be more detailed, and the lighting would have a much larger area to cover.

I imply above that the designers' reactions to my ideas would be simple tasks. At the time, I thought this was a simple idea. I didn't realize the extra work for all involved from the conductor and his orchestra to box office staff and house manager. In conjunction with that idea, I also saw a *passerelle* or runway extending into the audience.

Another HUGE decision was related to the space where I would direct the show. I had never worked in this theatre, but was always fascinated by its design. To give you a reference, two theatres share a single stage. This stage is built between two very different audience spaces. One side of the stage opens up to a huge 4,000-seat auditorium while the other side of the stage opens to a 700-seat proscenium where we would be performing this production. If you rent one theatre, you get the other one.

> R: Yes, we must RENT our theatres at our university. They are controlled by a commercial producing organization and union houses. Collaboration on a different level!

Before I reveal the HUGE decision mentioned above, I need to describe the climax of the show. It takes place on a dock where Queen (the main character) and her problematic lover go when they dream of the future outside of the life. They refer to it as their "special place." We never see this location until the climactic scene, but Queen describes the dock earlier in the show:

> *Queen*: He (Fleetwood) can still be like he was. When I get him away from bad company. Just last Sunday he took me to this spot we go to up by the Hudson. All that water and suddenly one of them big liner ships goes sailin' out of the city. "Queen," he says, "someday we'll be like that boat. Goin' somewhere that ain't here."

This location had to be different from all other scenes in the show. As Queen prepares to escape the life (premise!), she recalls this location as a portal out of the trapped and smothering world of the crowded Times Square.

My visual idea for this scene was that the trapped and confining world of the rest of the show would disappear and we would see the expanse of the huge auditorium revealed after two and half hours of literal entrapment on the set. The huge void of the darkened other audience space behind the set would represent Queen's escape and the unknown promise of her life to come. Once I had discovered that one image, the show fell into place for me. Again, I made a decision directly related to the design of the show.

> *R:* Even as I write this, I wonder if those kinds of decisions demanded by a director are overstepping my collaborative bounds. Is that not my area? Is that too limiting? By wanting that moment, I have defined that scenery must be flown or movable. I have forced the designer into a certain aesthetic. Is that too much?
>
> *K:* I feel that a director has to make certain decisions. Otherwise we're all walking blindly into the production. "Limiting" is not the word, it is about focus and how we are all going to lead to the climax so the audience does not leave confused. Flying and movable scenery has been done for years. When you have a director who is as excited about this concept as Rob was, you want to explore that moment and see how the rest of the show falls into place. Is that too much? Not at all. If I had a similar or "better" idea and he would not listen, then yes.

I was ready to begin to think about the world in which these characters lived. It was now up to me to find ways to share this analysis and decisions with my designers to create a cohesive production that would play into the strengths and camouflage the identified traps of the script. Any production—in design, direction and acting—must address, celebrate, and conceal all.

I began by searching for images for the designers that would refine my limited technical vocabulary and funnel the production in the direction I envisioned. There are so many pictures of New York, prostitutes, and neon. Believe me. It was up to me to winnow those images to support my vision and spark the designers to create their elements.

This part of the process is an important step. Did I envision these prostitutes to be glamorous or garish or subtle? Were they wearing un-theatrical clothing or were they overdone caricatures? Did I see the set as a literal interpretation of New York or as constructivist and sculptural? Did I imagine neon and garish colors or harsh, realistic tones?

Sifting through the images, I found myself drawn to close-up photos of overflowing trashcans, spray paint tagged buildings, ill-fitting clothing on sickly women, and street lamps that cut harsh shadows across the street denizens' faces. Those images kept defining my style of the production. Hyperrealistic (Kirk actually is the one who coined that word for this production). I ultimately wanted grit and dirt and reality.

R: To be quite honest, I never worked in this way before. I always felt it was the designer's job to bring in the images. I would then point to those that I liked. But being relatively new to this institution, I wanted to try working in this new way. I prefer working like this because it really helps me spark discussion with the designer. On the other hand, I also don't enjoy it because I become married to an image and idea and the designer becomes an imitator of that.

K: I always like when a director can speak through words as well as images. For myself, I find it easier to work with a director who works in evocative images expressing the sensation or smell or taste of an environment rather than telling you that this building should be stage right. But in either instance, the description and the gestures he or she makes while talking about the imagery speaks more to me than the image itself. Any way to better understand the director's vision.

In choosing these images, my history played into their selection. I had been propositioned by and witnessed prostitutes while living in New York, and not a lot of glamorous ones. These were the women I wanted to see. I had seen the details on the buildings and watched the burnt-out bulbs of the porn theatres reflected on the wet street. I wanted the audience to see that.

I then began researching prostitution to get a better sense of the culture. I marveled at statistics of sexual abuse relating to the choice of going into the life. I also read articles and biographies of current and former prostitutes. I was fascinated by how they entered this profession and why they stayed. Each new bit of information helped me focus the style of the show.

While in this phase, I stumbled on an article online written by a former prostitute describing her experience seeing the Broadway production of *The Life*:

Popular entertainments like *The Life* take prostitution for granted, just as prostitutes do, while getting on with the story. *The Life* does not "experiment on" prostitutes, as a feminist vehicle would. Nor does it try to make broad feminist statements. Instead, prostitutes and pimps are treated the way other characters are treated in musicals.[1]

Her assessment of the non-exploitative nature of *The Life* was exactly what I wanted to capture. This subject must not be salacious, simply—life.

My next phase of research was viewing documentaries of real-life prostitutes. One particularly seminal moment was when the microphone was left on while a prostitute serviced her client in a car. I was fascinated as I listened to her hot and heavy language while making the sale. Her voice later turned into purring during the sexual encounter. As the meeting continued, she turned to frustration when the client couldn't climax fast enough. She had a business-like tone when asking for more money to finish him. After he agreed, she morphed back into the sweet talk that was in turn followed by a hasty goodbye as soon as he finished. This was eye-opening. I had always thought about the sale, but never the moments after. The awkwardness of the encounter and the actual act had always been secondary. I realized that just like any good actor, the prostitute has to play many roles. This would help my costume designer and actors as we created biographies for the characters.

I knew that I had to do more research. I was going to New York to gather images and interview real prostitutes (chronicled in chapter 2), and that would have an obvious impact on interpretation of the show. However, I was tired of working alone. I had been researching for months (it was now late June) and I wanted to share my work with others. My upcoming trip to New York ended right before the next academic year began and I couldn't hold off the discussion till then. So, once I had gleaned the precious imagery and background I needed, I began to assemble a presentation for my first design meeting.

I put together a collage of images, statistics, and ideas that would help me focus my discussion with the designers. The presentation was the time for me to streamline all of my research and discard anything I had learned that was not in keeping with this production. The nearly 100-slide presentation was designed to help me stay on track with this large show while simultaneously giving designers an idea of the scale of the production.

For the costume designer I found a rather hastily shot photo of a prostitute in handcuffs on her way to the police station. I was drawn to the obscene ordinariness of the image. The carefully chosen style of the outfit was probably selected to highlight the woman's assets. The innocence of the floral pattern in the clothing in concert with the exhausted physicality of the woman made me extremely sad.

For the scene designer I chose a close-up photo of a filthy hotel room that had a matted shag rug, a heaping butt-filled ashtray, and a spilled tumbler of liquor. I was drawn to the details in the image. I loved the stories that seemed to be happening outside the frame of the photo. The well-chosen details implied a continuation of the life. Who is smoking? What are they drinking? What are they waiting for? There is danger in this photo. A fire could start. Someone could be passed out. I want to know what is happening in this world. It is about the details.

For the lighting designer I chose a very hazy image of a neon-covered street with several indistinguishable people on their way to nowhere. I was drawn to the removed, distant quality of the lights in the photo. The people seemed to have no connection to the environment and no identity. I loved the harsh shadow and the nonreflective metal in the surroundings as well as the halo effect surrounding the people caused by the light. For me, the most striking aspect was that I knew exactly what time it was in this photo.

I was excited to share my work with the production team. My enthusiasm in setting up the meeting was not shared. It was summer. The scene designer (yes, Kirk) had just finished an arduous few months designing five shows and was exhausted. The costume designer had come off a very challenging summer of work as well. The lighting designer was still working at his summer gig. The choreographer was still not chosen. The musical director was out of the country. But ahead I went, full steam.

R: I would say this is one of my best and worst aspects as a director. I am so excited to get to work on the project that I usually have to push for early meetings and find myself disappointed by the lack of enthusiasm from my designers. It eventually works out, but initially I can be a pain.

K: Yes, you can. But it is that enthusiasm that initiates the excitement in the rest of us to jump into a production because, after all, this is why we do what we do.

As I prepared for the meeting, I was excited to truly begin this collaboration. We were about to work on a show I knew we all loved. Surely disaster could not strike!

In the next chapter, we will review the first production meeting, its aftermath, and the designer's research before the meeting.

Note

1. Available at http://www.salon.com/archives/1997/date05.html (May 30, 1997). Tracy Quan is a contributor to the anthology *Whores & Other Feminists* (Jill Nagel, ed. New York: Routledge, 1997) and a member of PONY (Prostitutes of New York).

CHAPTER 2

THE LIFE IN DESIGN

With our theoretical collaboration assured and many lofty goals in mind, we immediately began the collaborative process by breaking many tenets of the first part of the book. It seemed, even for us, that collaboration was about adaptation. We aimed for the perfect collaboration, but factors such as time, schedules, and other projects plagued many steps in the collaborative chain.
In this chapter, we will examine

- initial design discussions;
- adapting theory from the first part of the book;
- dealing with collaborators in different stages of preparation; and
- analyzing successes and failures of collaboration in the initial design phase.

The First Production Meeting

The first design meeting with director and designers was a bust. The lighting designer was in another state working on a show. The scene and costume designers seemed shell-shocked from their hectic summer schedules. Unsympathetically, the director plowed ahead, as he would be leaving town for three weeks. The timing was disastrous.

At this meeting, the slide show of prostitution images and statistics was shown. The director spoke of the details in the presentation and how they related to the separate design elements. The premise of the show and dramatic structure was covered. He also distributed some initial thoughts on the show, included here:

The life means different things to different characters.

For Queen, Sonja, and some of the hookers, the life is a degrading and humiliating, dead-end patterns of sex for survival. *Queen is still determined enough to escape.* Some of the hookers also still retain dreams outside of the life, while Sonja is resigned to her exhausting life of doing the only thing she knows how to do.

For others (mostly the men), the life is a promise of riches that comes from using people weaker than themselves. Jojo's greed, Fleetwood's drug-muted ambition, Memphis's wealth, and Mary's blatant ambition are all driving forces to gain or retain a higher place in the life.

In all cases, the life means a daily routine of selling oneself to get ahead. In all cases, the life is a dead-end.

The Life *Premise*

Escaping the life leads to sacrifice.

The premise relates to our lead, Queen, who is willing to sacrifice everything—her body, her intelligence, her pride—for the promise of an escape from the life. Her misdirected sacrifice to users such as Fleetwood and Memphis is a terrible mistake. It is not until she sacrifices her heart (her love for Fleetwood) that she gains the tools (strength and empowerment) to truly escape.

For Sonja, her sacrifice of taking the blame for Memphis' death may give her a chance to recuperate and escape the life.

For nearly all others, the inverse is true: they refuse to sacrifice. Mary, Jojo, Fleetwood, and Memphis ALL use others to gain a foothold in the life. They do not escape. Fleetwood ultimately does sacrifice himself at the climax of the show. It is rare in a show that only two characters are redeemable.

The inciting incident is the discovery of the money used by Fleetwood for drugs until the climax, when Fleetwood sacrifices for Queen. The play is about people unwilling to think of others.

Thoughts on the COSTUME Design for The Life

These women do not change very often; they don't have the time or money. If they are Memphis' women, he buys them one thing that shows off their assets, and that is their "costume" until it is threadbare. The costumes for these women tell their stories—either from their past or what they are currently experiencing. Their clothing and nicknames are synonymous—sad details that really evoke an entire character. I want these individual pimps and hookers to have individual personalities as well, without going into *Guys and Dolls* territory. These are colorful, garish, and cheap people without going into cliché. They DO want to attract attention.

As for Queen, she is a little above the rest without being rich; she is just in a different class or tries to be. At the hooker's ball, go over the top—this is what they save for all year. This is when they try to outdo one another (rent the documentary about this event). Don't be afraid to go really far with the competition to be "classy." The rest of the time these people are not the cleanest.

Thoughts on the SCENE Design for The Life

I keep coming back to the dead-end image. The life has no escape. They are trapped in a walled in block in New York with no chance of escape.

Can I have an Act Curtain of Chain Link and Big Dead-End Sign like in the Images I showed, or is that TOO CLICHÉ?

I can see it going several ways:

1. Is this a metaphorical interpretation of the dead-end and the city? With forced perspective, slanted walls, and bizarre angles that show us the messed-up world they live in?
2. Is it a (selectively) realistic approach to the city, with bricks, metal, and graffiti?
3. Is it a combination of the two?

No matter what you decide, I want the set to be overpowering. This is quite truly a dead-end, so no horizon peeks through. They are trapped. I see every location popping out of parts of the set (no visible scene-change crew). For example—Fleetwood's hotel room is a Murphy bed hidden in a wall. Perhaps the Dollhouse takes place behind a painted scrim, which we don't notice until that scene lights up (maybe doubling for Memphis' hallway). How about the balcony scene for "People Magazine" as a fire escape that can be pushed out from the wall to expand? These kinds of tricks are emblematic of their entrapment, which makes us and them oppressed by the world they seem to never escape... There needs to be many places to hide and overhear and surprising entrances. EXCEPT FOR the climatic scene at the dock. This scene must be like no other. I want as much of the life (e.g., buildings fly or pull away) to disappear as a possible offering of the one sliver of hope. It is not until they get down to the rat-infested docks that they can even see a ray of hope. The life of course returns for the last moments of the play.

I need a way for the audience to be involved. I want them assaulted. At intermission I will have Three-card Monte games and hookers mingling as the second act begins. I'd like to have a runway around the pit so that I can REALLY get people in their faces for numbers like "Check it Out" and "My Body." I'd like to try it and still keep it in the world of the play... so, a cement runway with trashcans?

Thoughts on the LIGHTING Design for The Life

I have talked to you a little bit about the use of neon throughout. This can help us in establishing locales such as Lacy's and the Dollhouse, but I also think it can be a wry comment on what is happening, let those neon signs comment on the locations, the emotional state of the characters inhabiting them...a flickering "Jesus Saves" during "You Can't Get to Heaven" or a check cashing sign during "Mr. Greed." The colors in New York at that time were very harsh and quite overpowering: blues, reds, and harsh white. The lights need to be very helpful in defining the space for each scene. If the set is realistic with windows, I want those windows to light up as if the life is continuing...maybe even some spots or specials coming from those windows for things like Queen's ballads (breakups and patterns). The lights are also confining us to the life except for the dock scene. Here, across the river (although it is logistically wrong as the river they would be facing is to the west), I want the sun to be peeking out for the first time in the play. I would like to give you a chance to show that. Experiment with it. I really want you to use the dark other auditorium space if you can (e.g., a searchlight shot from the back of the auditorium or lights coming from those balconies...what would that do?) I want that scene to give us some hope, but not a happy ending feel.

Thoughts on the MUSICAL DIRECTION for The Life

The show is about being in your face: selling oneself and assaulting the audience, so the vocals should match. That means singing (safely) the hell out of the numbers. Listen to how long those women hold the notes in "My Body!" To me that says, "We are passionate about this, we mean it, we're not kidding." So, to get a place to maintain healthy vocal production with a real sense of danger and affront is the challenge ahead.

Thoughts on the CHOREOGRAPHY for The Life

I do not see traditional musical theatre choreography. I see a more modern influence of individual patterns that perhaps come together at moments, otherwise these people are distorted. They are not doing kick lines, they are selling themselves and trying to gain a foothold over the other. So, think odd levels, weird angles, and character-related poses à la Michael Chekhov's psychological gestures. There are times when a number may sound traditional (e.g., "Easy Money" or "People Magazine"—but look at what they are singing about...it is off-kilter go in that direction).

Following the director's inundation of base information, there was minimal discussion. The costume designer had one question (in two parts)— "How many people in the cast and how many costume changes?" The scene designer asked about the overall feel and what type of movement there was in the show. The director secretly resented their (seeming) lack of enthusiasm.

The meeting ended with apologies from the designers for being ill-prepared and the director for pushing. The next meeting was scheduled for a week before classes began in August. The designers were charged with finding production-related imagery while the director headed to New York City to research and get some more concrete images for the designers.

THINGS HANDLED INCORRECTLY

This meeting should not really have taken place at this point. It needed to take place in the spring before all were drained from their summer work. Without key players—such as the choreographer, musical director, and lighting designer—the meeting was fruitless. It was anti-collaboration. The director came in with too many ideas and suggestions. The script was not discussed in concert with the designers.

THINGS HANDLED CORRECTLY

The initial thoughts from the director on design and show elements were quite nice. It does seem to veer much too closely to making design decisions before they had a chance to breathe (e.g., the Murphy bed).

Further Research

Rob headed to New York with a camera and some gumption. He walked the streets of New York searching for vestiges of the old Forty-second Street—not so easy since Disney now owns the area. He found the right iconic images and architecture on the less-traveled streets of Fortieth and Forty-third Streets. There you could still see remnants of the dirtier time in the city's life. Even though most of those buildings were now being transformed or razed they had the right look for the show. Those images would serve the scene designer well (See figure 3a).

To aid in his own process, Rob began to spend time with some of the prostitutes working in the early morning near Forty-second Street.

> *R:* At first I was a little shy and just hung around the periphery of their work area…observing. More than once I was stared down by a mistrusting pimp. At one point, I went over to the ladies and to see if I could ask them a few questions. This prompted one of them to yell "He's a cop!" I started to run away, terrified, yelling, "I'm a not a cop. I'm a director!" Over the

(a)

(b)

Figure 3   (a) Director photos from New York City research trip
(*Source*: Photo by Rob Roznowski). (b) Scene designer thumbnail
sketch (*Source*: Scene Design by Kirk Domer).

next several days, I gained the trust of these women and observed their business first-hand. I gained a lot of insight from the story of their lives. This was invaluable for my costume designer and my actors. Some of the costumes chosen for the show and some of the story lines for the smaller characters in our production came from this experience.

Rob then met with original cast members from the Broadway production and contacted Ira Gasman, co-author of the books and lyrics, to gain more information about the creation of this show. He also made a terrible mistake. He went to Lincoln Center to watch the taped version of the original production. We say this is terrible because of the nearly impossible chore of erasing those choices from one's mind when working on a new production. The decisions made by another always seem to be simmering in the back of one's mind when making decisions on the new production.

Finding Balance in the Team

The designers, on the other hand, continued their mad pace of working on other shows and had made little progress when the next meeting was scheduled in late August. At this second design meeting all designers were present, along with the stage manager. Two key players (choreographer and musical director) were still missing.

This August meeting included a much livelier discussion that kept pushing everyone to define a mood, a tone, a style. There was much talk about the very realistic nature of the piece that still must support breaking into song. Discussions abounded about the multiple locations and the oppressive environment that acted as another character in the show. After sifting through the director's photographs and looking for details that interested each collaborator, the style of the show emerged: hyperrealism. Although no one could easily define it in the traditional way, everyone understood in this context.

R: I think that decision came from our discussion of the photos that I had taken and Kirk's desire to actually use them in the set. This collage of images from real locales would be hyperreal. More than iconic photos, they became tangible elements in the environment.

K: To me, defining hyperrealism had a lot to do with the intermingling of the photographic imagery with found objects and materials. It was as if we were trying to integrate the "stink" and cluttered world that was New York during this time, while maintaining the musical theatre conventions set up by this show. Hyperrealism is the overstimulated environment where whores might sing out loud and dance in unison.

With the director giving much background information on the production, the designers were now ready to discuss their contributions. They did so in dissecting the show and asking practical questions. This step proved their understanding of the script while the director monitored the discussion, noting which parts of the show most interested or concerned the designers. This sort of script discussion gave everyone a clear understanding of the parameters of the show.

The designers still proved to be at a standstill. The costume designer said she could not sketch until we had cast the show because of the very personal nature of the costumes and the actors' bodies. The lighting designer said he had images, but really wanted to see what the set would entail before going any further, even though he had researched neon lighting and had found some cost-effective alternatives. The scene designer said he would call the director tomorrow. At the next production meeting all players promised to present detailed research.

THINGS HANDLED INCORRECTLY

The director should have held back and let his designers make more decisions. By having such a firm hand, the director limited further design ideas. The designers should have been more forthcoming with research images.

THINGS HANDLED CORRECTLY

The discussion of the show and its strengths and weaknesses was exactly what needed to be taking place (perhaps a few months earlier).

Reworking Issues

The scene designer called the director as promised the next day. And then came one of the most calamitous decisions for the production team. Even armed with the knowledge of the chronicled experience for this book, Rob and Kirk had a separate meeting to discuss the set without the other designers. This meeting, which lasted several hours, resulted in a very fruitful experience for those two that totally eclipsed any collaboration with the rest of the team. Even though the other designers stated that their work could not begin until casting or scene design was complete, they should have been privy to the main part of the discussion, not simply the result.

During this meeting, Rob discussed the movement of the show and how each scene flowed into the next. Kirk paid careful attention to Rob's hand gestures. By intuiting what Rob imagined, Kirk began to ask questions and raised points, which snowballed into a collaboration duet. The free-flowing exchange of ideas and blurring of roles between director and designer happened exactly as described in the introduction of this book. It is that sort of energy one seeks when collaborating.

The following are notes from Rob and Kirk after this design meeting:

Director Notes

Went slowly through photos taken by director for designer; asked specific design questions to assist in finding the right elements for the show.

Director is stuck on a few images: porno theatre marquee, Murphy bed, and the climax at the dock.

Director described major transitions in show and his subconscious hand motions created the movement of the show.

Elaborate discussion of the Hudson pier. How does it reveal? Director keeps saying it is a safe place, yet, as designer noted, the safest place has most danger in entire show.

Choreographer must let us know how much dance space is needed for numbers.

Environmental qualities of premise is very clear as this is a dead end.

Prologue of show is in a different year when the area is being bulldozed, how to do that? Flashback in time? Discussion of first moment when set is revealed.

The first implication of AIDS hinted at end of show. Research needed to set exact year.

Designer Notes

Designer asked director about the interior vs. exterior world of the play. Why, with sculptural set (if that is the way to go), are there realistic interiors? There is no refuge.

Special problems: Union house does not allow any adjustment to playing space when director wants environmental.

Discussion of how the city is overwhelming. Designer worried that it may become too complicated to distinguish real icons; rather, suggested sections of these icons.

One musical number has several locales. How best to achieve that on unit set?

The Hooker's Ball—How elaborate? How to get into it? Is it always there? Large introduction to number leads to silent section—how to solve that?

Shifting of streets. Differentiation of streets. Does Lacy's have a sign? Is there a hotel sign? When we go to Dollhouse is there a sign? Answers: YES. The addition of a marquee sign ties it all together.
Designer pressed for character information, like Jojo—is he redeemable? Director is thrilled that designer wants to discuss character motivations.

That evening Rob and Kirk planned to meet in a few days to discuss more ideas following an intense round of research from Kirk, who would arm himself with answers to some of his major questions. More specifically, he would discover the parameters in order to guide his quest to design Times Square in 1983.

> R: I recall this day as a really exciting part of the process. Do you think we should have included (and probably bored) the others with our set-heavy discussion?
>
> K: This was the productive part of the design process for me. I felt that it was completely necessary for future discussions. It could have been help-ful to the other designers—this is true, but there are times when I feel each designer should have a one-on-one with the director in order to be productive. Design meetings can easily shift to needless discussion with individual agendas that are not productive.
>
> R: Looking back on the notes from it, we really discussed more than set; we got into some really great character motivation questions and intricacies of relationships within the show that all could share in.
>
> K: We were truly creating the world in which these people lived. Even though it was a real place in a real time period—this was our place in time and we could not divorce the inhabitants from the environment. If we did, we would have completely ruined the show and the hookers would have moved down the block.

Specific Issues

At their next meeting, the discussion of the major elements of the show was solidified, and details were now the order of business. Four major locales were needed:

1. The street—look for the majority of the show;
2. Lacy's Bar—where the extended book scenes took place;
3. Fleetwood's motel—later doubling as the upscale bedroom of Memphis; and
4. The climactic scene at the dock.

This covered the majority of the show, but certain smaller scenes needed to be addressed, such as the elevator scene, the jail cell, and the Dollhouse

Strip Club. How would these smaller necessities impinge upon the larger scenic elements? The designer would spend the meeting trying to decide which way to go with this design from the options offered by the director in his notes to the designers—realism or sculptural?

Notes from the second meeting of Rob and Kirk follow:

Designer brought other images that were more abstract and artistic rather than literal. Got ideas of perspective and what both agreed sculptural was. Director chose several favorites.

Discussions of the space needed for each major playing area led to a discussion of traffic flow and seamless transitions. Working together, going through the script, looking at each transition to make scene changes fluid. Certain scenic elements fell into place naturally to make these changes happen.

Major premise discussion of climactic dock scene and transition into this very important scene. Both agreed the set should "disappear" for this expansive scene, but how would the previous set restore for the end of the show? Designer saw it as act curtain coming down to hide the restore so that the crew could do it. Director saw it as another seamless transition. That moment was debated and collaborators returned to the idea of premise and the lack of escape. The restore would be seamless.

Designer became director:

Designer came up with a directorial decision that street would be divided into two distinct sections—stage right as the business side of the street with the bar and marquee, and stage left as the residential area of the show with the apartment and motel room.

Designer discussed traffic patterns for certain scenes and created blocking.

Designer suggested Queen singing onstage in jail while Mary and Fleetwood are still visible in the motel room making love. (This ended up as one of the most powerful directorial elements of the show.)

Director became designer:

Director offered solutions for set placement to tell the story more efficiently.

Director decided on unique elements of set to keep the design cohesive.

Director asked designer to make a quick sketch. Designer drew a very rough idea of what had been discussed. Director asked questions.

The director then asked for a quick ground plan. While he could certainly understand the quick sketch, he needed to see it from above to help him understand the world. Another quick sketch of the ground plan followed. Both seemed to agree they were on to something.

Two major design elements were left unanswered as the meeting ended:

1. Where would the jail be?
2. In the Hooker's Ball sequence, at the end of Act One, what would happen?

Questions posed by collaborators as the meeting ended:

> Does designer feel director has too much control? Did director design? Does director feel designer has pushed the show into a new place? Does designer question concept? Does director feel that wrong decisions happened with quick answers? Does designer agree with all the answers? Should designer not ask so many questions in order to have more freedom? This initial phase of collaboration is about making instant decisions that affect the whole, so the director and designer should know a show inside and out.

The meeting ended with a momentum spurred by the drawing. A simple line drawing offered enough fodder to springboard the design for the entire show. Both director and designer left feeling cheered and excited to see what would happen next.

THINGS HANDLED INCORRECTLY

As stated, the other design collaborators were left out of the debate on the parameters of the show. The director had many preconceived ideas about certain elements.

THINGS HANDLED CORRECTLY

The discussion of the major elements of the first day led to a detail-oriented second meeting. The director and designer switched roles during this phase, exactly as hoped. The tone of respect and trust in these meetings led to fruitful debates and finally answers.

Reactions and Revisions

Following this flurry of ideas, the next day the designer sent a quick thumbnail of the set (See figure 3b).

The director wrote down his reactions to the set along with many questions:

> First viewing of sketch of the set:

> I was disappointed by the realistic aspects. Designer warned me of the *Guys and Dolls/Porgy and Bess* elements. And that is exactly how it looked.

It had none of the grit or sculptural/architectural elements of the best parts of the research images. How do you best phrase that disappointment?

I took a long time examining it and looked for the positive aspects. I really didn't like much of it as it had a very "Universal back lot" feel.

I was silent for quite a long time with obvious disappointment registering on my face and this book ringing in my ear.

So, I first looked for the elements I DID like. There was a cage for the jail on a building that really made no sense. It was that feel of this bizarre element that I liked.

There was a bridge structure that jutted out and had no end that was more in keeping with the concept. I went to those.

I also recognized the ground plan was workable.

Then I had to discuss the elements that I did not like. It all looked too musical comedy.

It had a very clean and sterile feel. I felt my heart sink.

Is this person I am working with understanding me? Am I unrealistic? Is he working outside of his element? How to make that work? Is there a way to bridge that gap?

Designer assured me that in order to "mess it up" he had to start with the realistic and the logical. Did I ask to see it too soon? Was I pushing for that sketch too quickly? Is there ANY way to get out of this or am I just going to settle for some sort of facsimile of this original design?

The blunt reaction was handled well by the designer, who had already prepped the director for the first viewing. The director (although warned) did not seem to be understanding of the process. Following this awkward discussion, the designer and director prepared for the next scheduled design meeting to share their work on the production.

Inspiring Collaboration

The next design meeting included all of the key players. The scene designer presented his initial thumbnail sketch and explained the "trick elements" of the set. The lighting designer instantly became active as did the choreographer since they were greatly impacted by the scenic decisions. The costume designer began to discuss color palettes within this cityscape. The collaboration was sparked into action by one designer taking the lead in creating the world.

R: I know this process in the initial stages was problematic for the designers because of other commitments, but why do you think the scene designer is more often than not the prime shaper or leader of the design?

K: I find that sometimes the environment is the first thing that the director thinks about. Deciding what the place looks like impacts style and composition for the rest of the show. It is with the initial sketches of the scene designer that the director can help determine the scale, the reality, the tone of the entire world. Other times this process is led by the costume designer, lighting designer or sound designer. In this case it simply happened this way.

Several important discussions began after showing this thumbnail sketch. The lighting designer was concerned about where his lights could be placed on the towering massive set. The scene designer showed him several places where he had already addressed that concern. The choreographer wondered about the levels, steps, and ramps for the dancing. These practical questions gave way to a discussion of how to make all of the elements cohesive in this production.

The designers began talking about how hyperrealism must affect every choice. The designers spoke about the weird height perspective of the upper buildings. With that kind of skewed perspective, how can costumes mimic that style? The doors and frames were all normal, but the rest of the higher set became forced perspective. They agreed that the style of each choice in the show must match. If the set is skewed, then costumes, hairstyles, choreography, and acting must all correspond. All must be working on the same project. The director sat back with glee as the designers all scrambled to work together in harmony.

THINGS HANDLED INCORRECTLY

The other designers were ambushed with a first draft of a set.

THINGS HANDLED CORRECTLY

The discussion of matching styles is what every collaborator dreams of. The director sitting back as the designers worked their magic was excellent collaboration.

Expanding Designs

The next design meeting was what all had hoped for. The designers were all prepared and enthusiastic about the project. The director was simply offering guidance since the process was running smoothly.

The costume designer was interested in the mixed materials on the set. Texture and layers for costuming was discussed. Just as the set was brick, steel, and cement, the costumes would mix materials such as leather pants and a fur jacket.

The costumes for the thirty-member cast depended upon the individual character backgrounds. What assets are the girls selling? What does a pimp's clothing reveal? These sorts of questions could only be answered after rehearsals began. A questionnaire was made up in consultation with the costume designer that explored the background of each character. Questions such as "What body part are you most proud of?", "What makes you special amongst the other prostitutes?", and "What are your ultimate goals in life?" would flesh out the very specific needs of each character's costume.

The lighting designer began to play around with different looks for each scene in the show. He seemed to discover what he must do in order to support the premise, while offering unique looks on the unit set. He began to discuss how his neon and practical signs could enhance the mood for each scene.

The scene designer had addressed the director's concerns with his initial thumbnail by offering this Photoshop version of the set, using most of the director's actual photographic research (See figure 4).

The director was impressed that the concerns had been addressed and was excited to begin the next phase of details and revisions. He wrote his reactions down:

> Relief! Excitement! This is not *Guys and Dolls*. The elements are there in a hyperrealism.
> The photographic quality of the design is exciting.
> The city overwhelms and dwarfs the people as it swallows them up. There seems to be no escape. The designer has made a very interesting, complicated, and energized design.

Figure 4   Scene designer digital rendering
Source: Scene Design by Kirk Domer.

I love it!

My concern is that we must maintain this quality and not lose it to traditional painting techniques. Other practical placement of certain areas must also be addressed.

THINGS HANDLED INCORRECTLY

Nothing.

THINGS HANDLED CORRECTLY

At this point collaboration had truly begun and each member of the production team was working as hoped.

The designers and director felt confident to head into the next phase of the collaboration—casting and production meetings. While all felt enthused by taming this beast of a show, there were several elements that had been neglected or brushed aside in order to design the "meat"' of *The Life*. This oversight would result in several problems.

CHAPTER 3

THE LIFE IN REHEARSAL

W ith a solid base established for the show, the production meetings and rehearsals began. The entire production would eventually involve over 100 cast, crew, and musicians. To keep everyone in the loop, production reports and e-mails abounded. This chapter contains snippets of e-mails, notes, and other *Life*-centric communications.

This chapter will examine

- design elements once rehearsal begun;
- miscommunications;
- specific collaborative problems and solutions;
- production meeting challenges; and
- exhaustion.

Production meetings began while the primary collaboration took place in collusion during hallway discussions and separate meetings. Production meetings and rehearsal reports now seemed to address only the largest managerial issues of the show while the important factors of the design were solved in private. The director took immediate steps to keep all involved by creating a breakdown of the major dramatic action of the play and the collaborator's goals and their ideas for all scenes. Although each collaborator had a separate area in the chart, the collaboration extended beyond the borders. It was important that all collaborators knew what had been discussed or decided. This breakdown was made for each of the twenty-four scenes in the show. Here are two examples from Act One.

Scene One

Action of Scene: Time: 2000, 7:30 pm and then following the prologue 1983 1:00 am. Jojo begins in audience. Police siren and lights start show. When musical pulse starts the world begins to come alive. Action for beginning of song takes place behind act curtain. First people with lines in song escape through fence during "Check it Out." It all erupts as act curtain flies out. Action spills into audience. At end of "Use What You Got" we flash to the beginning of the main action of the play. Queen enters through house. Is there some action down in front of pit? "Lovely Day to Be out of Jail" Queen believes that her new life begins today. Smooth transition as Queen ends song as she heads onto stage and into Fleetwood's. Stage Manager: I mention sound throughout in other categories, can you compile?

Set: Audience enters through police barriers in lobby. Begin with set in the "present" (2000). Our unit set is ready to be destroyed to make way for a Coldstone Ice Cream Shop. Act curtain goes up in "Check it Out" after "Are you lookin' for a good time?" and next "Check it Outs" on musical build.

Costumes: Jojo should look older at beginning. Successful in the porn business now. Perhaps a coat removed to reveal his "hustler" outfit. All cast appear in basic whore costumes. Mary is in her post-Dollhouse outfit. (Change before her next entrance to "Midwest" outfit.)

Lights: Preset is shadowy and dark and does not reveal much of the set. Police light and siren mark beginning of show. It is hard to discern as audience enters. Jojo in spot for his monologue? During "Check it Out" street comes to life. Moments of introduction of each cast in specials? At end of "Use" Jojo is alone and the play begins. Lights need to help us go into a "lovely day" (one of the few daylight scenes in show) for Queen's scene and song. Most of song is in audience and around pit.

Choreography: "Check it Out" is first assault. We bring full cast out through entrances and doors and into audience. Must come to stage for intros of group or character. "Check it out" is intro to style of show and comes from each person's character. "Use What You Got" is the theme of the show: Use others. Depending on casting of Jojo (he is the ultimate expression of the greed and manipulation) he may interact with all through the dance. Each character uses another. "Use What You Got" should be REALLY choreographed to set dance vocabulary of show. End of song all disappear but Jojo.

Music: "Check it Out" is sung from character. Some are aggressive; some demand the audience to come to them. "Use What You Got' is slick. The feeling is singing from a place of no conflict. This is how WE operate. Jojo's vocals are effortless (yet desperate) that dichotomy is how each person communicates. In keeping with rawness of the show and exposed lighting, can we keep pit exposed by taking down the curtain. Any thoughts? Will it be a problem acoustically?

Scene Four

Action of Scene: Time 5:00 pm. Lacy's Bar scene is HUGE! We see Sonja and the other women for the first time. We see Memphis and feel his power. We see Fleetwood's relationship to Memphis and see how important Mary is to the show and Jojo's manipulation. And for the first REAL time we see Queen is really her own worst enemy.

Set: Lacy's Bar is part of the street. It is a place to regroup for the girls. It is a place to scheme about business for the men. Need to have a separate area for Mary and Fleetwood from Memphis and rest of the group.

Costumes: We see the girls for the first time. Memphis is the "best" or most conspicuously dressed of all.

Lights: Lacy's Bar is in the day, but the bar is always dank. Do we see light from outside entrance in this scene?

Choreography: Director stages all numbers here.

Music: "Oldest Profession" is a real one act play. Let's see the arc of this song. By the end we should love Sonja. She is tired...can vocals match that? Are those riffs a cry for help or exhaustion? (They are not simply to show off.) Memphis' amazing voice comes from a self-satisfied excess of power.

Vocabulary Issues

You may think that because the collaborative team all came from the same educational institution that vocabulary would not be an issue. One of the first and most major communication snafus came with the word "passerelle." Rob kept talking about a way to get the action closer to the audience—using the word "runway." Kirk said, "You mean a passerelle." Rob not wanting to appear ignorant said, "Sure."

They went round and round discussing the passerelle for the show when in fact all Rob desired were ramps to bridge the action into the audience. Miscommunication kept this aspect of the design from moving forward as the collaborators argued about budget and the aesthetic of a pageant type runway looping around the orchestra.

In the end, simple ramps bridged the action between audience and actors.

Auditions

As mentioned earlier (in chapter 1), this show was a co-production with the College of Music at our university. This implied a sharing of responsibilities and talent. The Department of Theatre would provide the design and crew while the College of Music would provide the orchestra.

The musical director for the production was at home in the symphonic world and fears abounded that he would be difficult to work with when asked to lose some of his orchestra pit due to design constraints. By approaching the situation from the start with a clear design idea as well as several alternatives, the maestro was more than happy to accommodate the request, since he had been part of the collaboration. All fears about his classical background were obliterated as he often donned a pimp hat to conduct the show.

Students from both theatre and music would be eligible for casting, and, as auditions grew closer, fear amongst the collaborators about the shared casting pool was evident. Workshops were held for both areas to prepare the students for auditions and both areas were eager to be involved in such an undertaking.

At auditions as well as the casting session, representatives from both areas were extremely polite and supportive, resulting in a cast of equally distributed leading and ensemble roles among the two areas. Again, a collaborative worry had been aborted through lack of ego, respect, and unhidden agendas.

Casting and Costume

With casting complete, the costume designer was eager to get started. Armed with documentary, visual research and biographies completed by the cast, she began to create a panoply of costumes that both revealed and enhanced the work of the ensemble members. Her work also showcased the best (and worst) of the women's bodies, the hierarchy of the pimps and hustlers, and added depth to the lead actors. Her initial renderings revealed a sensitive understanding to the piece and to the overall design of the production (see figure 5a).

Designers at Rehearsals

The lighting designer now came to the fore and seemed up to the challenge. As the only student member of the production team, he had the most to prove and the most pressure. The following is an excerpt from an e-mail exchange between director and designer:

> Dear Shannon [Lighting Designer],
> I have solidified a lot of things with the set this past weekend in terms of its movement, its timing, and its transitions. I would now like to start moving forward in other areas.
> I find these production meetings little more than a chance to catch each other up on certain decisions made during private meetings. I don't like that, but they are way too overcrowded to make any design decisions at them. If you designers want to meet outside of meetings (with or without me) simply let me know.
> Please schedule a time for us to go through all transitions. See if Karen [Costume Designer] wants to be there.

In case you didn't notice, I am trying to keep things moving and get as much decided as possible before rehearsals begin to get crazy for me and I become focused on a whole new set of challenges.

Did that mixture of materials and mediums give you any ideas for design decisions? I guess you were already there with neon and practicals, eh? Maybe there is more to be found?

Thanks,
Rob

*   *   *

Hey Rob:
There a few things that I want to let you know. Just so you know, as well as to ease your mind, as soon as the model is finished and the blocking pattern overview you mentioned is done, my role is going to pick up more. I have a lot of things in my head right now. They can't be concrete yet, but they will be soon. I wish I could take a picture of what is in my head right now for you, but trust me this is going to work and be awesome. As far as your comment about just designers meeting, I think that is the only choice we have with this show. I would like to observe even little meetings that you and Kirk have about the set or that Karen and you have about the costumes—observe, not say anything, but observe. Most of my ideas come spontaneously, and the more exposure I have, the more I can sift through later. It's going to be great Rob. This is by far the coolest set and all around best show I have designed. I'm not letting you, Kirk, or even myself down. Thanks and talk to you later.

Shannon

The lighting designer clearly understood the director's initial concerns and asked to be collaborative in a different way. His way of working seemed to be as a reactionary artist. While he is spontaneous, the director seemed to want everything solved and on paper.

These disparate collaborative styles were solved by two major decisions. First, the lighting designer sent several renderings of possible light looks for the show and discussed it with the director. This allayed the director's fears immediately (see figure 5b).

Second, the lighting designer began to be a regular attendee at rehearsals, with a separate table to notate while watching choreography and scene work. He began to create extensive cue sheets with his own interpretation regarding what the director or choreographer had said.

Cue: 21

Time: 12

Page/Line: Bottom 18, Fleet "I feel like a pinball…"

Figure 5 **(a) Costume designer renderings** (*Source*: Costume Design by Karen Kangas-Preston). **(b) Lighting designer concept renderings** (*Source*: Lighting Design by Shannon Schweitzer).

Effect: Just have the window light on Fleetwood to show how pathetic the situation is

Cue: 22

Time: 5

Page/Line: Top of 19, "I wanna piece..."

Effect: He's @#$ up now. Quick cue...almost a bump, daytime w/ breakups and a bit of color

Challenges

While all seemed to be going along swimmingly, there were a few brewing battles, which would compromise the design and the production were they not handled immediately.

In a show of this scope and budget, the production facilities were pushed to their limit. This massive show in the midst of a crowded production schedule made for some very exhausted colleagues. The technical director informed the designers and director that some of the set may have to be rethought in order to complete the build.

Immediately, the director and scene designer wanted to discover what could be done. Following consultation with the technical director, ways to simplify the set were offered by all parties. Immediately, thoughts returned to more of the sculptural, found-object idea. This would still maintain the shape of the set (necessary due to blocking and choreography completed), but could more easily simplify build issues. Rethinking the design to accommodate the work force was certainly a priority. In the end, the technical director agreed to complete the project as designed, but the issue was real and tested the limits of the technical students and faculty.

Choreography Complications

The next issue that upset smooth sailing was the choreographer. The student choreographer was doing exceptional and detailed work. This was not standard musical theatre choreography, but rather exactly what the director had asked for—a very modern and athletic approach. The issue was her time-management skills. Choreography rehearsals eclipsed scene and music rehearsals and the show was losing momentum in order to facilitate her thoroughness.

Several options were offered to make this part of the collaboration smoother. First, the director asked her to approach the musical numbers in larger chunks, mapping out sections and then going back to refine the details. This seemed to work for a while, but the details still slowed everything down.

Next, the director asked the choreographer to teach "assistants" the movements to speed up the process. The assistants ended up being the director, stage manager, and musical director. At one point, all three were in separate areas of the rehearsal hall working with the dancers. That was a collaborative sight.

Finally, separate scene and choreography rehearsals were scheduled in order to complete this element and keep the production moving smoothly.

Realizing the Design

Another element of the collaboration that was thought to be solved reared its head in rehearsals. The marquee to the porn theatre (which the director was adamant about) ended up being too small to stage the scene and song. What looked massive in ground plans and renderings proved cramped and unusable. Only one major scene took place in this space, but it was a pivotal moment and the blocking and staging could consist of little more than "Inch your way stage left while she tries to counter you." This problem should have been solved in design discussions:

R: Why didn't we catch that?
K: We knew this was going to be an issue as the original design was cantilevered too far to be stable without having support underneath. After discussing this we chose to sacrifice this scene a bit for the rest of the show.
R: Nothing could be done to make this scene and song work and each time we got to this section of the show I cringed, but the idea was great!

Changes through Rehearsal

Despite these problems, the show proceeded smoothly toward technical rehearsals. The production meetings continued weekly until the show's opening. The thorough notes from these meetings (prepared by the stage manager) indicate a very harmonious and respectful group working on the same production. While design decisions never seemed to occur in these production meetings, they were a place for all to catch up on the decisions made separately. As evidenced in the meetings, those decisions fit cohesively in the production along with timing and deadlines.

One major element of the show that seemed to morph throughout the production meetings was the song at the end of Act One called "Hooker's Ball." This number is intended to show off the hookers in their finery at this annual event. It is really a competition for the hookers showcasing their most lavish or outrageous outfits. It is a turning point for the main character and she leaves her cheating boyfriend for a more powerful man. The ten-minute sequence only contains one dramatic action—Queen chooses Memphis over Fleetwood. This major sequence seemed to always be on the back burner for everyone including director, choreographer, and designers. Below is how the production notes record this ever-changing, Act One closer.

- 9/2 SET NOTES: The Hooker's Ball will take place on the street. Things will be brought in to make it glitter. Elements of the set will draw light and make it pop. COSTUME NOTES: The Hooker's Ball will be themed, but not sure how yet. Mardi Gras? Personal themed?
- 9/9 COSTUME NOTES: The Hooker's Ball will be a Drag Queen Mardi Gras theme. LIGHTING NOTES: The Hooker's Ball should be tacky Christmas lights, a mirror ball that flies in, area lighting, and should look "weird."
- 9/23 SET NOTES: There will be tables brought on for the Hooker's Ball. They will be down right and left. They will be bought on by the hustlers. COSTUME NOTES: For the costume changes into "Hooker's Ball," the first to exit must be the first to enter to make the changes work.
- 10/6 SET NOTES: There still needs to be a decision made about the "swag" for the Hooker's Ball. COSTUME NOTES: Queen's gown for the ball will be very glamorous and "Oscar-esque." The timing for entrances and changes is 2:47.
- 10/6 MAKE-UP NOTES: We will need to decide at a later date who gets elaborate make-up for the Hooker's Ball."
- 10/20 COSTUME NOTES: The character of Silky may have an African feel to his Hooker's Ball costume. It depends on what the other pimps are wearing. The hustlers are not in the Hooker's Ball.
- 10/27 SET NOTES: The mirror ball for Hooker's Ball will be up center. The Hooker's Ball sign will be a mixture of flashing lights and mirrors. Kirk, Shannon, and Rob need to discuss the sign after the meeting. COSTUME NOTES: Jojo will be getting dressed onstage for Hooker's Ball. All Costumes are glamorous versions of the characters' original costumes.
- 11/3 COSTUME NOTES: Jojo will not change onstage for the Hooker's Ball.

The majority of the meeting notes reveal a production that is on track with all making choices quite definitively and quickly (considering the scale and timeframe of the production). However, it is quite obvious from an impartial reading of the above that this one element of the show was never solidified. Going into technical rehearsals with this issue still unsolved proved problematic and continued to haunt the collaborators.

Technical Rehearsals

The technical rehearsals for such an endeavor progressed beautifully. Each new element was introduced and the collaboration worked smashingly. It was a textbook example (actually this textbook) of what good planning and excellent support can foster. Each new design element complemented the next, and, despite fatigue, all involved were proud of their elements separately and the show as a whole... until Monday evening.

The following is a journal excerpt from the director regarding technical rehearsals:

> Monday's rehearsal was the introduction of the orchestra and microphones. Once the orchestra began playing my heart stopped. I never heard another word from the stage for the next 2 hrs. The balance was off: the music too loud, the microphones feeding back, and actors shouting to be heard over underscoring! When I agreed to working in this barn of a theatre, I demanded better sound equipment for this production. My face was ashen for two hours as I was not able to hear a word. I went into panic mode, but I never lost my temper. It's just so important to keep calm in moments like this when you want to scream or yell at someone.

> Luckily the collaborator from the College of Music had worked in this space often and had ideas. It was so great to delegate responsibility to somebody who had worked in that space, which is notoriously acoustically problematic.

> Many adjustments were made to the space and to the sound equipment to accommodate all people in the production. Tuesday was spent running around campus trying to find curtains for sound proofing and repositioning huge speakers. The schedule for Tuesday night's technical rehearsal was abandoned and the entire evening was given over to deal with sound issues. The actors and designers were in great places, so they all agreed to this sound-centric rehearsal. A production team has to be brave enough to work outside of the rehearsal schedule to solve unforeseen major issues like that.

Rest assured the sound issue was solved by the scene and lighting designer working with the musical director. This issue had been averted, but several others hovered over the heads of the production team right up until opening night and beyond!

Problem Areas

Hooker's Ball

The costume designer spent a large portion of her budget and time on extravagant costumes which appeared onstage for less than ten minutes. Her costumes were quite successful here. Other elements in that section were not.

The Hooker's Ball locale had gone from outside on the street to tacky hotel ballroom and finally located in a nondescript place. This setting was to be capped by a large sign proclaiming "Hooker's Ball." The sign (an afterthought really) was created haphazardly after a trip to the local Home Depot by the director and designer. Once the sign was complete and hung late in the technical rehearsals, another issue arose. No one could see it because the lighting designer had no instruments left as this sign never appeared in a rendering. This was a pure moment of communication breakdown on everyone's part.

Queen's Escape Costume

Up until opening night the designer and director were unhappy with the several possible outfits that Queen might wear for the last part of Act Two. At this point in the show she is "disguised" so that she may make an escape from the life. Numerous outfits were tried, but none seemed to work. The final result was a long skirt, which seemed problematic at best to make an escape.

The Sun

The big moment when the dock and the huge expanse of the other audience space were revealed never achieved the scope the team hoped. The moment was lovely, but could have been breathtaking.

Memphis' Bedroom

Literally three hours before opening night, the director and scene designer were spraying adhesive on aluminum tiles, to the headboard in order to make this bedroom "reveal" something special. In the first act the bedroom is located in a fleabag hotel. In the second act the same set is redressed as a luxurious (if tacky) overly decorated bedroom as Memphis introduces Queen to a whole new life. Talks of mirrored ceilings and animal prints were abandoned due to exhaustion and budget. The result was something less than glamorous.

Successes

Although the previous section of this chapter seems to focus on the moments of stress, compromise, or fatigue, the successes of the production must also be chronicled. The collaborative team worked together extremely well and created a production of which all were proud. That rarity in theatre is something that must be celebrated but also analyzed. In the next chapter of part II, the collaborative post mortem takes place among key players of *The Life*.

CHAPTER 4

THE LIFE IN REVIEW

As mentioned in the first part of the book, there are several ways to monitor the success or failings of collaboration. Those modes of examination hold the true test of a successful production. The answers are sometimes painful to hear, but offer the theatrical artist a true guide to better work in the future.

In this chapter we will

- examine reactions of the authors;
- examine reaction of peers;
- examine the reaction of the reviewer; and
- provide an unexpurgated conversation among the key players of the production team.

Reflections

The following is a transcript of a conversation that occurred immediately following the closing of the production of *The Life*:

R: We've never worked together as designer and director. What has been the biggest problem with that? Biggest challenge?

K: Well, I don't think that at any time we really lost our focus—because neither of us wanted to allow the other to do that. But at the same time, we made our already ridiculous schedule more difficult.

R: You've talked many times about my micro-management of things. Was that issue evident?

K: Yes and no. Did I feel challenged? Yes. But I felt like this was a true test of my collaborative abilities because of your very specific needs and visions. What ended up on that stage was my version of your ideas, not just a facilitation.

R: Agreed. I think corralling so many different ways in which people work with so many members on the production team made for a difficult process throughout. A challenge.

K: You were always asking the right questions, although there was some compromise in the design due to educational and time limitations. I had to cut some of my favorite elements of the set (the buildings at forty-five-degree angles, which would jut into the stage, just hanging but not positioned) due to time.

R: Those were my favorites as well, but there just comes a point when other issues become priorities. I think the fact that this is an educational facility really plays a major factor in compromises throughout any project.

K: I still hate that several little details or moments of the show were not solved in my area.

R: I must say that after seeing the final show last night I was excited and moved by things that I was not expecting to be moved by. We've been focusing so much on the technical, that I was like, "Oh, there's a show here."

K: Biggest surprise for me in the process was the costume design in relation to the look of the overall show. It was a surprise. It came very late in the process in my mind and just sort of fit with all else. It's responsive collaboration. I don't think it is ideal, but it worked for me on this production.

R: The biggest surprise for me was believing that I had thought everything through, and then elements arrived and surprised me. One would be the marquee, which is really small. Another would be the dock. We built this whole big thing for Queen to have this great exit, which worked just fine, and then, when it came into the space I had never noticed this two-foot wall that made it impossible for this clean escape.

K: Even the most organized production has surprises when you enter the theatre.

As this conversation occurred immediately following the closing of the production several issues must be factored into the analysis. The immediate rush of euphoria that accompanies even the most cynical of artists must be acknowledged. Also, OUR need to present the best face possible to future readers also affects the analysis of the event. Finally, pride in one's achievements makes even the most sober judgment somewhat skewed. With all of those factors in mind, Rob and Kirk again discuss their work on this production after a year's worth of events, experience, and education.

R: I look back on this experience, having directed five productions since *The Life,* as a very important one for me as collaborator. I have always been confident in my work with actors, but a little lacking in discussion with designers. I feel that my aesthetic always influences each production. I feel that MY taste always somehow ends up onstage. I hate that.

What was unique about this show was that, because of its scope, I was unable to micromanage each element—even though I tried. I see that I was hoping to handle each element of the production as we put this book together. I see the need to make sure that I am more respectful of the designers.

Was this a successful collaboration? I would say absolutely, with caveats. I think everyone on this production team worked beyond what they believed their potential was. I think we all continue to be proud of our achievements, but we all had moments we wished we had handled something a little differently. I think that regret of minor issues will be the case on any collaboration, but what was special about this one is that those moments recede while pride in and respect for your collaborators is what we now recall.

K: I look back on this experience, having designed twelve productions since *The Life,* as an experience that I have not seen since. I design a lot in operetta where many productions I work on are steeped in enhancing the tradition of the genre. Many times I find myself, and the small group of directors with whom I work, falling into comfort zones. We know each other's vocabulary and facilitate each other's needs so easily that we lose some of the spontaneity of the project. I hate that.

The Life was a project that was spontaneity at it best. What with working with a new director, another unit inside our university, and a production at least twice as large as anything we had attempted in the past—every day was a new challenge. All of that and we were trying to create the utopian collaborative experience ... what were we thinking?

Reflecting on the whole, I would have to say that we could have planned a little better. YES, we needed to start earlier—not only for the collaboration but for the educational value of all involved. We were trying something new and everyone met the challenges of the day, but what if we placed the bar a little higher? What if we had met the challenge— accepted it—and had time to reflect on it ... better yet, enhance it? That could have been something really special. But what am I saying ... I still sing along to the CD in the car and smile at every reflection.

Response from Collaborators

The next place one may seek feedback is from one's peers. Since our data gathering did not include unknown audience members (as stated in the first part of the book), we looked to our collaborators from a different discipline. Our peers in the College of Music were asked to fill out questionnaires following the run of *The Life*. We have edited their responses merely for space rather than content. The responders are Melanie Helton (*M*, Vocal Coach and College of Music Consultant) and Raphael Jimenez (*RA*, Musical Director and Conductor)

How would you describe the collaborative process on The Life?

M: Pretty much smooth as silk. Decisions were made with consultation amongst all interested parties; casting was a very creative process with equal representation from both areas.

RA: The collaborative process of *The Life* was excellent. It was the consequence of a very healthy relation between colleges based in the mutual respect of each other's area.

Do you think that you were an equal member of the creative team?

M: As much as my time and other activities would allow. I certainly was kept in the loop on all matters.
RA: Absolutely!

What are some memories of successful collaborative moments on that show?

M: Casting, first and foremost. Rob's and my comprehensive knowledge of our students work habits, talent and discipline were extremely useful in casting both principals and ensemble members. Our constant enthusiasm for the project was also a big bonus. Seeing my music students dance up a storm was a wonderful sight. I know they all learned a lot of basic stagecraft, as well as adding to their movement repertoire. Listening to the singers lift the musical level of the show was also great. As I was not in that many rehearsals hearing from my students how well the two groups were getting on and helping each other was also great.
RA: There are countless moments of great cooperation, but I particularly remember a moment in which I felt the necessity to express an artistic opinion on somebody's work. Not only was my respectful comment well received, it opened the opportunity to exchange impressions on each other's area, nurturing the artistic result of the show.

What are some memories of unsuccessful collaborative moments on that show?

M: I think most of the unsuccessful moments came as a result of the rather fluid sense of time and schedule. Students were over-burdened. Also, the problem of providing consistent rehearsal accompanists was constant and extremely expensive in the long run. The time issue with choreography also slowed the process, but in the long run looked great.
RA: We had only minor moments of tension due to the lack of control of the rehearsal time.
M: It seems to be that in every collaborative process like this, the distribution of rehearsal time becomes the main source of potential tension between the team.

What would you do differently or the same in future collaborations?

M: Joint casting needs to be continued, with all parties' opinions being respected. The students felt the casting was a very fair process, and that should continue. I'd also like to see if our music students could be considered to help in the various technical enterprises. As many of them are going to end up teaching in high schools, some technical skills would be invaluable.

RA: This team is a great example of professionalism, leadership, artistic integrity, and respect. I think everybody who participated in this project was motivated to perform at his/her best as a result of an excellent rehearsal environment.

Reviews

Another arbiter of the production experience is the reviewer. As mentioned, a reviewer is an excellent way to examine an individual's response to viewing a production. What follows is a review printed in *Lansing City Pulse* newspaper, of Lansing, Michigan, on November 18, 2005:

MSU dares to live "The Life" to the fullest

By TOM HELMA

It's a euphemism, expressed with a sense of irony. "The Life" is a sarcastic way of describing the life of a street prostitute, which is, of course, no life at all. Is it possible that a renowned jazz pianist, Cy Coleman, better known for the successful light-hearted Broadway musicals he has written, can create an operatic play of the same name—"The Life"— without also inadvertently glorifying the street life of a prostitute? As presented by Michigan State University's Theatre Department collaborating with the MSU School of Music, the answer is a resounding yes! Energetically danced and sung, of course, but not without a corresponding pathos for the despair and emptiness of the street lives of the characters presented in this play.

A cyclone-fence-caged pre-1980s 42nd Street and Seventh Avenue New York City street scene is reconstructed in the abstract by scene designer Kirk Domer. The message is already clear: This street is a prison.

The prostitutes, pimps and hustlers who live here are self-condemned to a life of desperation and pain. A neon light flickers on, the multi-colored torso of a naked woman blinking in reds and greens. Bar lights and the old yellowing bulbs of the Triple XXX movie theater also spread light across this run-down city block stage set as Frank Williams, in one of six lead roles, invites us to get to know Jojo, a beguilingly charming drug dealer and all-around hustler who leads the entire cast of well over 30 individuals in a hell-of-a-knock-out opening number. Wow!

Costumes are simply magnificent, with Karen Kangas-Preston inventing 18 very original slinky, skimpy, Fredericks-of-Hollywood style outfits. Tiger-striped tights compete with a lavender ruffled blouse for one young "thang," while one of the pimps combines purple velour with a red pork-pie hat.

The music is powerful; Brad Fowler's lead trumpet and Christopher Gherman's back-up trumpet yelp and screech over as many as eighteen other instruments delivering a tightly written musical score.

This play belongs, however, to Sharriese Hamilton, who, as Queen (the "ho" with a heart), alone escapes "the life." Hamilton, as Queen, is elegant and holds on to a sense of personal integrity, even as she is violently abused and beaten on stage by Memphis, the white-suited pimp played by Christopher Austreng.

Right beside Hamilton, however, is Bonique Johnson, a tall Amazonian actress who is just a bit too attractive to play the part of the over-the-hill mother-figure prostitute Sonja, but effective nevertheless.

Vocally, these two women have perfected the chanteuse style of jazz singing popularized by Dinah Washington and Billie Holliday and end the play with a touching duet about friendship in the trenches of lost souls.

Austreng, whose powerful bass singing voice reverberates throughout the auditorium, is a thoroughly effective villain, while Nathaniel Nose, as the love-interest pimp/boyfriend of Queen, delivers a solid performance as the embittered Vietnam vet turned coke addict.

Set designer Kirk Domer has saved his best contribution to this well-performed play for last. The New York street scene opens in the middle to reveal the blinking lights of an impossibly far-away New Jersey skyline. It takes several minutes for one to realize that he has taken advantage of Fairchild Theatre's having no fixed back wall, but instead only a curtain that separates it from the lengthy MSU Auditorium.

The female ensemble in this production contributes exuberance and enthusiasm to their roles, but at no time does one succumb to thinking there is glamour in the lives of these tragically limited people. These characters portray young women and men who likely were sexually abused as children, who, running away, attempting to escape one prison, find themselves now in another not very much larger one.

Bravo to Michigan State University for selecting this controversial work and performing it, in every respect, exceedingly well.

Post Mortem

What follows is a discussion between the main collaborators of the production team: Rob Roznowski—*R* (director), Kirk Domer—*K* (scene designer), Karen Kangas-Preston—*KP* (costume designer), Shannon Schweitzer—*S* (lighting designer), and Jessie Cole—*J* (stage manager). Although all were aware that their conversations may be published, all involved aimed for truth.

> *R*: What was your general impression of the collaboration as a whole? Go. Honestly.

S: Well, early on, it seemed as though, you [Rob], Karen, and Kirk had already had meetings and that you guys had your idea of what the show was going to be. When it came to me, you guys already had this term, "hyperrealism," and I didn't really understand what that term meant. I think it was the fact that the term had the word realism in it...and I knew it wasn't quite realism, but what did the word, "hyper" mean? At first I didn't really know how to make my lighting go with that term. For it to be a show that would work, I would have to come up with this "hyperrealism" lighting, and since I didn't really know what the term meant, I think that was hard for me, early on.

R: And I think the fact that we had already met is really important...such a problem early on.

KP: Well, I was at that first meeting, so I didn't feel that way, but knowing that you weren't there, you missed out on a lot of discussion.

S: Yeah, and I really think that that was part of the reason why, especially in this early production meetings that we had, I was very quiet. I wasn't necessarily bringing research to the table because Rob already had a concept. He had like thirty-some images and he was like, "Watch this movie." At first it felt like I was going to have to recreate something for this director that I've never had to work with before.

R: I think that was important too. I think that, what I have to do in the future is show imagery that isn't literal. I learned.

S: I think it does work to have more emotional images than actual images.

R: I agree.

K: I had that moment with you [Rob] when you could imagine a Hooker's Ball that I could not fathom. In my head...in my world...I couldn't. You [Karen and Rob] defined it as an extension of their regular characters. I needed something more because I couldn't figure it out. So it was kind of nice because you two just solved it.

KP: I had to though. I couldn't let that go.

R: And all of those discussions really led us to defining each person. And defining the color scheme and all of that. Making decisions like, *Do they have the same kind of influence?* and *Are they going Mardi Gras?* Those were really great discussions, I thought.

KP: Yes, this whole number involved a new sort of research. I had to research going through costume storage instead of looking at images. It took two minutes instead of spending half the budget and all of my time on it.

S: It's interesting that you left the Hooker's Ball for the end. Because whenever I'm doing lighting, I usually look at what scenes are going to be the most difficult and what scenes need the most work, and I usually do those first.

KP: But we saw so much of the characters in their everyday costumes that Hooker's Ball was secondary to that. It wasn't until we had the actors and until they had those character studies that they did—which were fabulous—that they were able to form those people. This helped so much because by that point we had already done a lot of the designing of it, and there were things I made changes to because of what the actors said. It was like, "Wow, if that's the way you see your character, I can relate to that, and I can shift something to make it work for them." Their character studies then answered how each character would dress up for the Hooker's Ball. It was the biggest part of the build, and we probably wound up spending the most on it because it was such a big, fantastical thing. But, it was still secondary to who those people were.

S: I came to Rob at one point because I didn't know what he wanted. At first we weren't speaking the same language...at all.

K: Isn't that the worst thing in the world?

S: Yeah. But then I went and I did those renderings. I basically created the same looks with all of those different colors. And I said, "What colors do you like and what colors don't you like?" That was the first step for me.

R: I'd never seen renderings like that so I had a real learning curve. I recall so many decisions that came from those renderings: footlights and angles of light.

S: The moment I felt like I fit in with hyperrealism was when we started talking about the angles. We didn't want to go so far out of this world, but we at least wanted the non-realistic scenes to look realistic, and we also wanted to break it up.

R: All of those types of angles were the ones that we agreed upon and circled. And right, that's the moment that you *did* become part of that world. We discussed those renderings and both agreed "Oh my God! That's perfect."

I've been involved with many other productions where something got in the way, and I don't think that ever happened with us.

KP: Not with this group.

K: We had to walk the line of friends or colleagues or whatever it was in this group. We were working with students.

KP: I still feel like we had a comfortable relationship. Especially within our group, the students are so much a part of us. It's not a separate group. So there's a comfortable relationship. And there's the fact that we knew each other. I felt very comfortable with this group because I felt like we were friends before we started working together. I knew we could have real conversations and that if I told you "No," that sometimes it was in fun, but sometimes it was, "I honestly have to say no to that." And you wouldn't blow up or not understand it. So I don't know if that was better or worse, but I felt comfortable coming into it.

R: But I had never worked with any of you.

KP: That didn't bother me.

R: But it did me, because I didn't know how any of you worked— including Kirk—

K: What was it like for you, Shannon, as a student?

S: Well, I've worked with you [Kirk], design-wise, probably for about four or five years now. So, there's something about working with him, that kind of makes me work a little bit harder. He always tells me: "Don't screw up."

R: The book's subtitle: Don't screw up.

S: But I know it means something about his trust that I'm going to do a decent job with the show. So working with him was not a problem. And Karen…Karen…I love Karen. Karen was great. The whole student thing on this…the only thing I was worried about was working with Rob. I don't know if it was necessarily me being a student. I don't look at it in terms of being a student. It doesn't really matter where I am, whether it's here or working in opera or some garage. In this show, there were times I was working in a whole new way. That happens every time, I guess.

R: So, basically what you're saying is that each show requires its own set of design tools.

KP: That's absolutely true. I research differently, I sketch differently, I write differently—it totally depends on the show.

S: The only thing that's standard for me is my light plot. I have to have a light plot and a channel hook-up. As far as any of the other pre-production stuff, or how I mentally engage myself in a show; it's not just how I feel at the moment, but it's also who I'm working with. Isn't that true? Like the best way to talk to a director: some directors want to be told things and other directors want to see things. Other directors tell you things, and those are the shows that kind of suck. Because then it's like, all right, fine, I'll be your facilitator.

K: The way you said it was beautiful. I go through the same thing. I work digitally. With this production I still had to work digitally in my mind. But, I ultimately had to give him a white model.

KP: If I'm designing a realistic serious piece or if I'm designing a huge comedy, I'm going to sketch to fit the show.

R: Right. And the research thing. I mean, you research differently for the style of the show, correct?

K: I still research the same way; I just look for different stuff. Should I do a rendering? Should I do paint elevations? Whatever works.

S: Other than the costume designer doing renderings, I think that the director needs more visuals from the designer in preproduction than anyone else.

K: I think in this show, for me, it was all iconic imagery. I rendered differently than normal, but then I had to go back and do a white model because my rendering lied.

KP: You have the drafting where you have to be specific enough to build off your work. I don't. I don't draw the patterns. I buy them. I make them, whatever.

K: Right. This was one of the hardest ground plans I've worked on in a while. The first ground plan gave me a hell of a time. It took me hours to just find out what functioned. I couldn't give more space to the marquee or you would not see 75 percent of the show. I told a lot of lies in my renderings. It was more about the feeling of the show.

J: On the whole, you all worked really well together. There was a lot of discussion outside the meetings, but during the meetings, it was all restated. It might have been repetitive to some, but it helped because everyone heard it and it never seemed like anyone was lost. Every aspect was touched on, and even though elements would be totally different week-to-week and there were sometimes huge changes, things like that didn't matter because in the meetings, they came out. And if there **was** an extremely crucial change, then it had been communicated to the specific departments that it touched on, and meetings just served to catch up me.

S: I think if anyone was out of the loop a lot of the time, it was actually Jessie. Even though many of us would have side meetings, we'd relay the information to another person. Even before the production meetings, we'd know what people had talked about during the week.

J: And that was fine, because once we got into the rehearsal process, the show was something that I saw every day and I was there for the crucial changes. Back in the planning process, it was just my job to make sure that—at some point—everyone was brought onto the same page. I think that having those weekly meetings did that.

KP: But you know, based on the sheer size of the show, I couldn't wait a week to ask a question. We had to talk outside of the meetings. And since we saw each other ten times a day, and we would ask somebody when we saw them, rather than waiting till a meeting.

R: We don't just see each other ten times a day. We made a *point* to see each other ten times a day.

KP: But we *could*. It's not like we were in separate cities. I'd call you, e-mail you, or see you, whatever. I'd work on whatever we had talked about that week at the production meeting until I had a question about it, and ask, so I'd know if I had to keep working or if I'd solved it.

R: And what was kind of cool about the production meetings was that there were always questions at the end of them that needed to be answered. I thought that was really productive.

KP: I liked that there were questions in the notes. It was like having assignments.

R: Yeah, and each of us had assignments.

J: The collaboration was just so great…you guys were so enthusiastically involved in the show that you made a point every day to communicate. When it came back to me, we were caught up. Even though there were ten new ideas, you had to come to a final conclusion.

S: Well, the four of us are very opinionated people, and the fact that it worked out as well as it did really surprised me. Like, when I went to all those rehearsals and was writing all those notes down, and Rob would come up to me and say, "Well, what do you think about this?" Down on my notes was exactly what you were thinking in your head, and we started comparing ideas that way.

R: And you gave me a cue sheet, and you went through—in your own vocabulary—what was going on in those scenes. It was beautifully said and I had no questions. So once you were cueing, it was really cool. You wanted to show certain things at specific moments, and it worked because you had experienced it.

S: By going to the rehearsals I was able to see how Rob moved the characters and the actors actually figuring out who their characters were. And for me—as a lighting designer—the fact that I went to two and a half weeks of rehearsals it really helped out a lot. I had never done it before.

One of the great parts of Rob as the director was that he was always available. He was very open to me sitting by him at all those rehearsals, where a lot of directors would have said, "I need to focus on the show." And there was time for going to look at the space and things like that. Rob didn't put anyone in front of anyone else on this production. I felt like everyone—stage managers, designers, choreographers—everyone was on the same level. I've worked on a lot of shows where a scene designer or costume designer was the main person on the show. As a lighting designer, you can feel that way. And...I do sometimes, but it was nice to feel like everyone was on the same page.

R: Where did that come from. The trust grew during the process? I didn't have that at the beginning until everybody started contributing great work. It was mostly me in those early meetings...'cause I'm pretty much a micromanager, if we all agree on that? Agreed?

K: No comment.

R: I mean, it's really kind of true. But don't you feel that, as the process went on, that I relaxed? Just a little bit? Karen, you have to admit.

KP: Around us.

K: You did not relax around us; you relaxed toward us.

KP: With us, decisions had been made, and we were in agreement, but there was a lot to deal with.

K: Right, but you didn't relax.

R: Well, I never relax.

K: No! I mean, you did. *To us*. But, we all felt the frustrations in all areas of the show because we were all invested. We all felt those things...because we cared.

KP: We were part of a team.

CHAPTER 5

THE LIFE IN PICTURES

Figure 6   Opening number "Use What You Got" shows the seedier Times Square coming to life

Source: Photo by Kirk Domer.

Figure 7   The Murphy bed in the fleabag motel and the Act Two "glamorous" Memphis' apartment

Source: Actors featured: Sharriese Hamilton, Nathaniel Nose, and Chris Austreng. Photos by Kirk Domer.

Figure 8   Production photo of the number "My Body"

Source: Photo by Kirk Domer.

Figure 9   The marquee scene as Fleetwood tries to win Queen back. Jojo looks on from above

Source: Actors featured: Nathaniel Nose, Sharriese Hamilton, and Frank Williams. Photo by Kirk Domer.

PART III

COLLABORATION IN THE CLASSROOM

R: The idea of TEACHING collaboration is tricky. But you have done so to great success at Michigan State. Our students are really thrilled with the class.

K: I took a version of this class at University of Wisconsin/Madison called Collaborative Studio. It was a graduate-level course consisting of MFA scene, lighting, and costume designers as well as technical directors and directors. We would create four to eight designs per semester in collaborative teams of three to four people. The class was always team-taught with faculty advisors from the respective areas. It was a great opportunity to test the process in a somewhat safe setting that allowed us to learn about working with different people, personalities, and schedules under a unique director's vision. While Madison had a different composition of faculty, graduate students, and resources to develop this course, I modified it for MSU to enhance the education of teaching artists.

R: As observer of your class and reviewer of final projects, I was really amazed by how successfully everybody worked in and out of their areas of concentration. That must have been an exciting process and taken a lot of encouragement for the student to risk working in a new collaborative medium.

K: It was exciting to see how each student took on a new role in each production team. Designers as directors tried to fix all of the problems they had seen in their past productions while attempting to present a clear concept within a production. Directors as costume designers tried to serve the production without creating or overtaking the show's vision.

R: This next part of the book is about transferring collaboration to the classroom. Don't you hate when someone tells you how to teach? We won't do that in this book, right? But I do think that it takes a true collaborator to teach this class. Am I wrong?

K: The biggest challenge with this version of the class is that I am the only person teaching it. And I have to ask myself: am I trying to create the perfect director for a designer? Am I perpetuating bad protocol by having the designer respond to an unclear vision? Those questions have to be asked.

R: Agreed, but, in a class that examines the process, you can address those issues in there, true? Personally, I wish this was a required course for everyone I worked with, including myself when I was starting out. Aw hell, I need to be reminded of this as I enter each new show.

In the first two parts of this book we provided examples for successful partnerships and examined a true collaboration from inception to completion. Through theoretical musings and detailed accounts, you have gleaned YOUR opinion of what we wrote and what transpired. So, in this part of the book, we will assist you by offering ways to translate your discoveries and pitfalls into a classroom setting—either as student or teacher.

How can we take concepts or events and transform them into successful classroom teaching tools? This book hopes to address these ideas by offering a comprehensive overview of the course taught successfully here at our school and ways to assist you in preparing to teach or partake in Collaborative Studio. Everything from syllabus creation to sample projects will be offered to assist the instructor. Along with teaching ideas, we will, as always, provide an insider's view of personal successes and failures. Students may learn from experiences of past students and offer their teachers ideas for possible projects.

While this part will assist in practical lessons, the course can certainly be adapted. The course outlined in this next part is designed for a fifteen-week semester of work and includes only directors and designers. However, it can certainly be adapted for any design, directing, acting, or arts administration project. The larger lessons can be reduced to a one-time class experiment in any basic directing or design class. (See the exercise entitled "Cooperation versus Collaboration" later in chapter 1.)

The concept of collaboration is not solely a theatrical domain, so ideas and concepts of leadership, management, and sharing one's vision can certainly be used for nearly any subject taught. The theatrical idiom is what we know best. Imagine collaboration for bus drivers...it can be done.

Finally, we want to offer ideas to assist in the preparation of a course or lesson while giving some warning and encouragement. We in no way mean to imply that you must approach the lessons described as we would. As always, you should create your own style and method to best relay concepts you admire to your collaborators. Basically, the collaboration between students and teacher is its own partnership.

CHAPTER 1

PREPARING THE COLLABORATIVE CLASS

In this chapter we will examine

- assessing your students needs;
- assessing your limitations;
- choosing the scripts;
- time management;
- prepping the class;
- prepping the student; and
- faculty support.

Theory in the Classroom

Collaboration in the classroom? Can a concept unique and specific to production actually translate into a graded, unproduced class project? Can students learn the skills necessary to handle crisis, manage budget issues, and collaborate successfully while playing pretend? Does this experience make for better collaborative experiences? The answer, we think, is—*yes*.

The concept of collaboration began for most of us way back in science class when you and your partner were assigned a project. In most cases, one of the team members did a majority of the work while the team shared the grade; the Collaborative Studio ensures that students be held accountable for their contribution to the projects. Furthermore, all involved in the class will become aware when a student does not fully perform her or his part of the collaboration. It is quite obvious when one element of an otherwise cohesive production is lacking.

The only issue that goes unaddressed in this Collaborative Studio is the challenge raised when you have a collaborator who is a good talker with lousy follow-through. We've all experienced one of those on our team—all style and no substance. As an educator you feel disappointed that those students will work in the real world because they talk a great game but lack any sort of expertise while executing the design. Although this sample class (due to time and budget constraints) does not address the issue of poor execution, be certain that reputation will. If that does not satisfy you, why not expand the ideas that follow and translate them into an actual realized production examining EACH step of the process.

R: I had a designer for my first show in New York and he dazzled me in his preliminary work and actually met deadlines. I was shown sketches and was told to "relax" when asking what sort of shape his area was in. I was still so scared to make waves. This show was being built off site and in another STATE and so I never got to check on the progress. On load-in day, the set arrived. It was ugly as sin and did not match the sketches. When it was assembled, the main entrance for the door opened directly into the side theatre wall. The designer had no explanation of how that could have happened.

K: I worked with a director early in my career as an educator who excited me with central metaphors that would explore a script in a new way. He would poetically describe these images that excited and prompted me to go home and create and transform the idea into a setting. When presenting my sketch the following day, anticipating a positive response, the director told me the whole concept was scrapped for a new concept he read about in the *New York Times* that morning. This continued for several meetings. Sometimes it's better to wait.

The need for the class is evident. If we can help one hotshot undergraduate designer to funnel his brilliance for the good of a production or assist one unfocused graduate director to communicate her idea more clearly, the theatre world will be a better place. The classroom is the perfect place to dissect the collaboration while examining it in process. Most design or directing students will be in the midst of an actual production while taking the class, and believe us—while this class is being taught the faculty behaves better, the students delve more deeply into the work, and the experience for all is enriched.

Preparing the Course

Most can agree the idea of collaboration is wonderful in theory and when related to a realized production. The idea of teaching ways to effectively collaborate is novel for most institutions including our own. Through discussion, it was agreed that a class on collaboration should be taught. One problem: there was not a course in the catalogue under which to teach the class. So, we covered it with an open-ended practicum course. In the curriculum catalogue it is described as:

PROBLEMS IN DESIGN—Theatre 811: Approaches to design problems in costume, scenery, lighting, architecture, properties, or sound.

That broad description became narrower as syllabus preparation began.

The first place for us to begin with any syllabus is creating the course objectives. By defining the class goals, the syllabus would be easier to create. The course objectives were as follows:

- Improve research methods through exploration with regard to direct research into the time period surrounding the text as well as evocative imagery that aids in the communication of your ideas.
- Explore the historical genres, theatrical styles, and dramaturgical context related to the chosen texts.
- Develop an understanding of presentation styles and standards.
- Improve cooperation and communication skills as they relate to the collaborative process.
- Advance critical assessment skills principally as they relate to the evaluation of your process as well as that of your fellow collaborators.

The basic outline of the class is simple: students are assigned various positions on several productions in which they learn to collaborate. They are designated as director or designer using assigned scripts and hold mock production meetings leading to a design that seeks to create a cohesive production while maintaining individuality and respect for all involved.

Since our university does not have a degree focusing in directing, the class was made up of a conglomeration of students—master of arts candidates, MFA designers, MFA actors, and talented undergraduate designers, actors, and directors. What makes the class interesting for all involved is that, in most cases, the student will be working out of his or her safety area of focus. Sometimes the lighting designer will be the director for the project while the actor will be the costume designer and so on. By working outside of their chosen study, students learn to respect the nuances and dedication needed for each role on the collaborative team.

The particulars of the class will become clearer as you read further, but for now you get the gist of what was trying to be accomplished. Since you know the basic idea of the class, let's examine the classroom collaboration team.

Needs of the Students

How to best create a syllabus for a class covering the creative process? Each time you are about to begin a class the preparation is the most daunting. Nothing is worse than staring at a cursor ready to formulate the best way for YOU to chart a student's education. The onus is crushing. If you are like either of us, you procrastinate.

In some cases, the procrastination is really the gestation period. As long as you are discarding and refining your ideas—even while watching television—you are preparing. However, in syllabus creation for this class, you must look no further than your students. Let their needs inform the choices and aims of your syllabus. By looking at students' individual and collective strengths and weaknesses, most of your prep time can be focused on addressing their needs. What does that mean in the context of this class?

Let's just imagine that

- Student 1 has never lit a realistic interior.
- The shows that all of these students have been working on for the past few years have been contemporary.
- Student 2 only works in a small arena theatre and needs a show with a huge cast.
- The department hasn't done a musical solely designed by the students.
- Student 3 needs to work with silk painting.

With each of the above statements, your syllabus and class assignments have begun to take shape or at least have been narrowed. You can begin compiling possible titles for the class to work on. So, instead of choosing from every play in the entire world, you now are going to work on a realistic box-set proscenium musical that is set before 1890 with silk painted costumes. Good luck finding that, but you have a start!

Self-examination

Once you have examined the needs of the class, turn the attention to yourself. What can you learn during this class?

Can YOU be a better collaborator? In teaching this class, you want to make sure you demonstrate some of the principles in this book. You should think about the example you present to the students when in a collaborative situation yourself. If you got angry at last week's production meeting, and were witnessed by members of the class, you can always explain the extenuating circumstances. However, setting the tone for a clear and positive collaboration is important.

Next, examine YOUR choices in projects assigned in the class. If they are all similar or if they are last year's projects, you could instead opt to expand your scope while pushing your students. This is the most difficult thing for a teacher to do, but being thrown out of YOUR area of expertise sometimes makes your work (at least your preparation) and your empathy with your students' struggles better.

R: When asked my first semester at a new school to teach Classical Acting, I panicked at first and then had great fun learning and refining and experimenting with what I had an acquaintance with, but felt a little scared to approach.

K: When asked my first semester at a new school to teach lighting design, I panicked at first and then had great fun learning and refining and experimenting with what I had an acquaintance with, but felt a little scared to approach.

So, if your students in Collaborative Studio need to design a Russian constructivist turntable for a French neoclassical play, you might want to research that genre.

Expanding the Class

The first time the class was taught, it was composed only of designers. This was because of the fact that no one else wanted to enroll in a class called "Problems in Design." Consequently, all of the directors for the mock productions were faculty volunteers rather than "directors" chosen from the class. That glitch inhibited the process. Aside from the obvious intimidation factor, the students had already worked time and time again with the faculty. This was not necessarily a bad thing, but it did not create an environment for growth from new experience.

So, the following year, an effort was made to expand the class to include actors, directors, and more.

K: Designers simply working with designers would not allow for true collaboration. So, out I went to convince skeptics to join the class. By explaining the benefits of such a class, I shored up enough support from many groups to make the class worthwhile for all involved. With student directors on board, I had to realize my own limitations—that I am not a director. So, Rob volunteered to mentor the directors outside of class. Why stop there? Since each "production" was given a budget loosely based on the theatre space chosen by the team, why not bring in the costume shop supervisor to deal with faux budgets and design clarity? Or bring the TD for design implementation and cost analysis?

Selecting Scripts

After reviewing your students' needs as well as your own, it is time to start choosing projects for your Collaborative Studio.

K: I knew that I wanted to have the students focus on three very stylistically different plays. I wanted plays that were open to interpretation so that these designers could imagine many possibilities.

You may find that three shows are too few or too many for your collaborative studio. Expand or reduce your syllabus. We feel that three shows could be more realized by spending time refining details. By offering more scripts, you may discover that if a student doesn't connect with a certain script, the individual at least has others to choose from. Our take is that the course has a process-based focus.

When you are teaching a class on the collaborative process, it is important to make the students part of the creation of their learning experience. You should invite the students into a pre-class forum to ask a question of them—what do they perceive as their needs? In most cases, your expectations will match their response:

- The costume designer wants to design period clothing other than Shakespearean or Elizabethan costumes.
- The lighting designer is tired of re-creating yet another 1950s interior.
- The set designer yearns for some explosive post-modern environment.
- The director hopes to find a unique way to serve the script.

Using a list of titles pieced together from initial thoughts and adding suggestions from the students, you are ready to begin. The meeting continues as you begin the process of honing these choices for your syllabus. At this meeting, make sure that you have a large supply of scripts on hand, so that you and your students can peruse them. In our department, we met in the theatre library, so the scripts were readily available.

Our discussion led to a stampede to the shelves where the students poured over their favorite playwrights. Off the shelf came Caryl Churchill's *Cloud Nine*. Three of the twelve knew the show and animatedly explained it to the others. Peter Shaffer's *Amadeus*, known only to the class from the movie, was reexamined for its theatricality and design possibilities. Then there was the old standby, Molière's *Tartuffe*.

Of course, everybody wanted to do a musical. Everybody has her or his favorite. In discussing titles another need arose, the challenge to create dance-worthy costuming. So we began to narrow. Gilbert and Sullivan shows are too presentational. Bernstein, Laurents, and Sondheim's *West Side Story* was too New York–y. Rodgers and Hammerstein were too overdone. Coleman, Fields, and Simon's *Sweet Charity* was finally agreed upon for the highly stylized environments, dance worthy costumes, different locations, and, let's face it, "The Rhythm of Life."

   R: Don't you think anytime you give students ownership of the class, the work and dedication is enhanced?

K: Without a doubt. I mean, people had favorite plays, but they entertained each opinion. I think it is always better to gather options, but know when to guide them to something that may really challenge them.

Toward the end of the meeting, all had chosen the period show *Amadeus* and the musical *Sweet Charity*. Yet, everyone agreed we needed a show that was experimental—a show that would support "uninhibited theatrical expression." Right or wrong, they felt their choices were too location-specific and wanted to find a show that could be open to novel interpretation. The fact was that the entire group could not collaborate on a decision for a final show. They wanted a script where students could run wild with a highly conceptualized show.

K: Acting as director, I guided them to Sophie Treadwell's *Machinal*. I explained its amazing possibilities and its episodic nature, just as the other students had done with their selections. This show was on the initial list I made, and the support for this progressive, feminist show grew. At meeting's end, I asked them to trust me, and penciled in *Machinal*. Initially, they agreed, but still wanted to read it first before the syllabus was finalized.

Days later, *Machinal* was read and approved by the students. The scripts were set.

Organizing the Projects

Once you have chosen scripts, look for realistic project timing when creating the syllabus. Creating a timeframe for the creative process is difficult. Also, trying to promote a creative atmosphere with responses and feedback, while still providing time to be a creative artist, is a delicate balancing act. Students are not only creating projects, but are also trying to build their portfolios. In our course, time constraints allowed about five weeks of design time for each show. Extra time was allotted for the musical.

R: Estimating the timing of when a class will peak is the hardest part of writing a syllabus. How much time can a project realistically take? It all depends on the students, plus myself, and how we all work together. I have to guess. In an acting class, I find that I end up pushing my agenda to match that proposed and arbitrary date. Even though I chose the date, I force them to be ready to perform for a grade on January 18th! I imagine for this syllabus it is doubly hard.

K: Exactly! Everybody's creative process is unique. The projects and the production teams are all so different. You just do your best. There are times when teams are not ready to present, but, hey, they had the schedule and, just like in real life, there is a deadline.

Another factor in choosing the dates for product presentation was practical. Because the class was composed largely of MFA design students who were involved in every department production, it was realistic to work around the department schedule when finalizing dates for project showings. Although it is true that projects can always be expanded to fill the schedule, it is a shame to not allow enough time to really examine the process.

A great resource to include, if your faculty or organization is willing, is bringing in production teams from recently closed shows. This post mortem is held to trace the collaborative process. The team is gathered to discuss the collaborative process from beginning to end. These sessions are incredibly informative to students. To see a realized production and then hear the challenges a production team faced is an invaluable tool that will enhance this class immeasurably. It is up to the instructor to promote a frank discussion that focuses on the collaborative process.

Observing another team's presentations can be invigorating and nurture respect. By the time you choose the dates for the final presentations, each group will have witnessed versions of other designers and production teams. When students observe the pitfalls and successes of others, they ultimately design a more cohesive production. Even when the projects aren't as successful, the students still examine the process.

When choosing projects for students on the syllabus, it is important that students occasionally be assigned to areas other than their normal emphasis. The benefits are obvious, and well-rounded collaborators emerge from projects with newfound respect for each position on the production team.

Similarly, when students present their designs, each designer was encouraged to experiment differently depending on the show. They had to vary presentation styles throughout the course, depending on the director's preference as well as the best way to communicate the show's vision.

In deciding the order of the projects on the syllabus, shows should get more complex as the class proceeds. In our collaborative studio, *Machinal* was first because it was the most freeing. In the first project, students get to know each other without being overwhelmed by an extremely challenging production. They easily see each class member's (and future collaborator's) style. They then refine the process with *Amadeus* and finally tie it all together with *Sweet Charity*.

Once the deadlines are set and the process presentations have been arranged, you must create an opportunity to see the teams in action. It may be helpful to schedule time into the syllabus for collaborative observation. Be sure to incorporate opportunities to allow production groups to kibitz with other design teams working on the same show.

K: I borrowed the word and the concept of "charrette" from my lighting design professor Linda Essig at University of Wisconsin-Madison. The definition of a real charrette is a typically intense, multi-day meeting promoting joint ownership of a solution that attempts to diffuse traditional confrontation between parties. By adapting the charrette concept into the classroom, moments of inspiration and debate about the production could be witnessed by all. Charrettes were built into the syllabus and were held during class immediately following the directors' presentation of their concept and the initial discussion of the play. Teams were asked to hold discussion until all could observe their collaboration. In creating the syllabus, be sure to set up intermediate deadlines for directors and designers alike to react and respond openly with the design team in front of the class.

With all of the preparation completed, you are ready to finalize your syllabus. If you recall the first part of the book, we discussed vocabulary. So, to prepare you for reading the syllabus, here is Kirk's vocabulary lesson:

Collage rendering—a collection of images representing the final look of the show.

Color scheme—the color palette for each designer. They may be fabric swatches, paint chips, or gel colors.

Constructive and collaborative response—an opportunity for each member of the production team to question or comment on the presentations; the conversation is expanded to include all members of the class.

Costume plate—full-color rendering (by hand or digital) to represent the fashion of the character.

Director's mise en scène—setting the scene. How the director will communicate the world of the play through all of his or her ideas and nuances.

Dramaturgical support—can include background on production history, socio-economic factors of the play, the author's history, and historical aspects when the play was written and when it is set.

Full-color/digital renderings—perspective drawings in a painting medium or digital program (e.g., Photoshop, 3-D Studio Max, Renderworks, etc.).

Full-color model—a painted or digitally printed model representing true color in a three-dimensional form.

Light lab—using stage lighting and sample gels, this enables the lighting designer to refine color and intensity choices using costumes, fabrics, and scene painting samples with actual lighting instruments.

Light plot section and hook-up—lighting paperwork in which lighting designers communicate needs to a master electrician.

Paint elevations—full-color representations of sections of the design. They are usually presented in conjunction with a white model. These may be drawn by hand or digital.

Paper dolls—layered representation of costumes made from paper.

Post Mortem—A gathering of key players from past productions to discuss the successes and failures of the collaboration.

Research/mood board—a presentation display board representing direct visual imagery related to the show as well as evocative imagery representing the ideas of the presenter.

Rough ground plans and sections—quick sketches to define spatial relationships in regard to scenery and lighting placement in top view and side view.

Rough model—a preliminary three-dimensional communication tool, usually of foam core to roughly show spatial relationship, mass, shape, and form.

Schematic diagrams—drafted plans representing lighting direction, color, source, and feeling.

Script and character analysis—general analysis of the play that can include plot, time, characters, theme, premise, environment, imagistic language, physical needs, and dramatic structure.

Story board—a scene-by-scene pictorial of the entire play.

Thumbnail sketch—an expressive doodle that bridges the gap from research to initial design.

White model—a "no-color" representation in a three-dimensional form to define the world without color. This differs from a rough model as it is a final version of the design.

Sample Syllabus

Below is a basic overview of the fifteen-week, semester-long seminar class that was taught twice a week for two hours at each meeting.

Week 1

Class # 1 Introduction; Discussion of Syllabus and Course Expectations
        *Read: Syllabus
        *Read: *Machinal*
        *Bring Drawing Materials
Class # 2 Lecture: Cooperation vs. Collaboration (see Exercise #1 below)

Week 2

Class # 3 Discuss *Machinal*
Class # 4 *Machinal* Director Presentations
        *Bring Materials for Design Charrette

Week 3

Class # 5 Design Charrette for *Machinal*
Class # 6 *Machinal*: Scene and Costume Preliminary Presentations

Week 4

Class # 7 *Machinal*: Lighting Preliminary Presentations
Class # 8 *Machinal*: Light Lab Presentations
　　　　　　*Go See: Department of Theatre (Show #1)

Week 5

Class # 9 *Machinal*: Scene and Costume Final Presentations
　　　　　　*Read: *Amadeus*
Class # 10 Discuss *Amadeus*

Week 6

Class # 11 *Machinal*: Lighting Final Presentations
　　　　　　*Discuss Department of Theatre (Show #1)
　　　　　　(In-Class Presentations and Post Mortem with production team)
Class # 12 *Amadeus*: Director Presentations
　　　　　　*Bring Materials for Design Charrette

Week 7

Class # 13 Design Charrette for *Amadeus*
　　　　　　*Go See: Department of Theatre (Show #2)
Class # 14 *Amadeus*: Scene and Costume Preliminary Presentations

Week 8

Class # 15 *Amadeus*: Lighting Preliminary Presentations
Class # 16 *Amadeus*: Light Lab Presentations
　　　　　　*Discuss Department of Theatre (Show #2)
　　　　　　(In-Class Presentations and Post Mortem with production team)

Week 9

Class # 17 Collaboration Exercise (Voice and Movement)
Class # 18 Musical Theatre Conventions

Week 10

Class # 19 *Amadeus*: Scene and Costume Final Presentations
　　　　　　*Read: *Sweet Charity*
Class # 20 Discuss *Sweet Charity*

Week 11

Class # 21 *Amadeus*: Lighting Final Presentations
Class # 22 *Sweet Charity*: Director Presentation
        *Bring Materials for Design Charrette

Week 12

Class # 23 Design Charrette for *Sweet Charity*
Class # 24 *Sweet Charity*: Scene and Costume Preliminary Presentations

Week 13

Class # 25 *Sweet Charity*: Lighting Preliminary Presentations
Class # 26 Collaborative Analysis and Problems Discussion
        *Go See: Department of Theatre (Show #3)

Week 14

Class # 27 *Sweet Charity*: Light Lab Presentations
Class # 28 *Sweet Charity*: Scene and Costume Final Presentations

Week 15

Class # 29 Discussion of Exam Portfolio Presentation
        *Discuss Department of Theatre (Show #3)
        (In-Class Presentations and Post Mortem with produc-
        tion team)
Class # 30 *Sweet Charity*: Lighting Final Presentations

FINALS   GALLERY SHOWING: Each collaborative team presents
        (including the process) each show in a presentational gallery
        format.

Exercise: Cooperation versus Collaboration

The following is an exercise that may be used for ANY design or directing class as a one- day examination of the difference between cooperation
and collaboration in relation to working on a production team. Some form
of this exercise may also be a possible way to begin your Collaborative
Studio.

First, we asked the students to offer synonyms for the word "cooperate." At a recent seminar where this exercise was introduced, the words included:

- respect
- share
- partnership
- assist
- get along
- play nice

Then we asked the students for synonyms for the word "collaborate." At that same seminar, those words included:

- unified
- integrated
- melding
- meshing of ideas
- harmonious
- cooperate

You may note that "cooperate" was mentioned under the synonyms for collaborate. When pushed, the student who offered the word said, "[w]ell, you do all of those things under cooperate when you collaborate. Except you are working together."

Without a lot of discussion on the subtle differences of those synonyms, students were handed a short text. This text can be anything. We used free-style rap, but you could use beat poetry or straightforward old-fashioned drama. The students were then asked to read the text.

Once they explored and understood the text, they were then given a handout with an image in the middle of an empty stage. The images were different for each person—an ancient African jar, a shiny hubcap, a large pug dog, or an apple. These images were not chosen to go with or against the text, simply as random images.

Students were then told that the image in the center of the stage was the director's defining metaphor for the text that had just been read. Students were then asked to design the setting for the text/show incorporating that image. Instructions included that they may not alter that image in any way. Rather, they had to use that image in conjunction with what they analyzed from the text and create a design for the show.

The resulting designs were not successful—a cityscape with a giant African urn stage center, a large shiny hubcap surrounded by platforms, a pug dog with scaffolding, and an apple blocked by large bridges.

When asked to explain the designs, students kept coming back to the idea that they were working AROUND and AGAINST the image rather than with it. This central metaphor had no relationship to the meaning of the text and they found limited ways in order to incorporate this random image with their ideas. It is true that some designers liked having that image as a starting point for their design but still felt trapped when they could not alter its scale or coloring.

In the next step we asked the students, "If this image was handed to you in a production meeting and the director told you this was the central image for the show, what questions would you ask?" Immediately the students began asking the right kind of questions:

- "Does this African jar symbolize ancient tradition to you? Do you like the shape of it?"
- "What exactly about the hubcap did you like? The shiny quality? The coolness?"
- "That dog looks mean. Do you think the show is about being attacked?"
- "Is it the redness of the apple or its white core? Is it to be eaten or is it plastic? Is it the beginning of life?"

As the questions kept coming, the instructor was monitoring and answering these questions that in turn inspired more questions and discussion. These questions were helping the students refine and adapt their image. In this seminar, discussion about an image flowed from one student to the next. They found inspiration and many found connections to the text.

Next—Students were then given either the same image as before or a different image that was used by another. They must again create the setting for the play, as before, EXCEPT they may alter the image related to the previous discussion.

  R: OK, here we differ. I think the students should be given the same image that they began with so that they can chart the process and see how these new rules apply. Through class discussion their design alters.

  K: I believe that you should shake it up and give them a new image randomly. For one thing, they may not have liked the image originally given to them. But more importantly, they had an opportunity to see and hear about ALL of the images used in the first presentation phase, so in this round they are able to collaboratively create related to everyone's questions as well as their own.

Whatever way you choose, the students created this new setting on a blank piece of paper while abstracting the image using energies, movement, colors, feelings, or qualities to enhance the text. You will usually find that these designs are much freer and more creative because the designers had a chance to use the portions of the image that spoke to them. (It must be noted that this seminar included actors, non-theatre students, outside observers, fellow faculty, and designers.) No matter the make up of the class, the results were the same—the resulting design from the second style is a freer and deeper design.

Students were then asked to show their designs. The others were told to look at the image and the resulting design. The students then discussed the ways they saw the image abstracted in the product. Finally, the designer explained his choices in the final product related to the image.

At the end of the class, we returned to the synonyms on the board and discussed the two design styles. The students were asked to look again at the words that define cooperation. The first way of designing was cooperation—simply "assisting" someone with the image. The second was obviously collaboration where it was a "harmonious meshing of ideas." This practical exploration of cooperation versus collaboration is an excellent exercise to prepare students for Collaborative Studio.

CHAPTER 2

COLLABORATIVE STUDIO

B elow are representative productions from the collaborative studio class. You will note that premise statements were not utilized in these projects.

Project #1: *Machinal*

Mise en scène from director:

Machinal

By Sophie Treadwell

Setting: Anyplace, anytime

Cast:

The cast of forty-four characters will be played by twenty actors. The actors who play the roles of Young Woman, George H. Jones, Telephone Girl, and Mother will only play one role. The rest of the actors will play up to three roles each.

Mise en scène:

The visual representation of *Machinal* is a vice. It is mechanical, industrial, and heavy. It has straight lines and hard edges, and is brushed silver and textured black. All of these adjectives describe the play. The vice also keeps turning and turning until it eventually comes to its peak tension point and cannot go any further. This represents the Young Woman specifically, as she is forced to be a part of this world that keeps pressuring her and forcing her into society's ideal, until she hits her breaking point. She is being held in place, when she wants to be free. Throughout the play, as the vice tightens, she is burdened with more things that keep her in place—job, family, husband, and child.

The group scenes in the play will be almost choreographed, with characters moving in relation to one another, at times, almost in unison but not quite. Each character's movements will be economic and with purpose, but not so much so that it is overtly obvious to the audience. The Young Woman only attempts to get in the timing with everyone else, but never stays there long. The loosest scene is "Intimate," where both characters are free and unrestricted.

My production of *Machinal* smells like wet pavement. On most days, the smell is harsh, dirty, and unappealing. However, on the first day of spring after a long, cold winter, when the weather is warm and sunny after a rain, one welcomes the somehow sweet scent. This day is similar to the scene "Intimate," when the Young Woman tastes happiness and love for the first time, finding magic in a sinful environment.

Setting:

The setting will be inspired by the feel of the vice. Greys, brushed silvers, and blacks will be used predominately. It will also have a machine-like quality, possibly being a machine itself and working throughout the play. The set will not place the play in any specific time period; it will be universal. The setting, for the most part, will not be elaborately changed from scene to scene.

Costumes:

The costumes will have the feel of uniforms, but they will not be identical. Everyone, except the Young Woman, will be in similar dress, using colors from the same family and tone, and they will be hard, definite lines. The Young Woman will wear a softer dress, and possibly a softer color, to represent her out-of-place feeling in society. Each person will wear an armband with a barcode on it.

Environmental Facts:

Geographical location:

Any smaller big city (i.e., Philadelphia) in America, but a city with seasons

Time:

Anytime past, present, or future; the seasons and time of day change for each scene

Economic environment:

All of the characters are working class, except George Jones who is middle-class. The city is, on average, a working-class and middle-class city.

Polar attitudes:

Young Woman: sees her world as very small and confining, knowing that the world outside her life offers so much more
George H. Jones: sees the world as his oyster
Mother: sees the world as traditional in values and expectations
Man (Young Woman's lover): sees the world without responsibility and full of women.

Dialogue:

The sound of the dialogue is like an old style typewriter: rhythmic and mechanical, except for the Young Woman who sometimes falls into this category but cannot stick with it. "Intimate" is also not in this category. The dialogue there will be smooth, comfortable, and more natural

Dramatic Action:

"To Business"	Workers attempt to impress.
	Jones moves in.
	Telephone Girl plays.
	Young Woman avoids.
"At Home"	Young Woman seeks rescue.
	Mother nags.
"Honeymoon"	Young Woman avoids.
	Jones convinces.
"Maternal"	Young Woman avoids.
	Jones cheerleads.
	Nurse protects.
	Doctor belittles.
"Prohibited"	Telephone Girl flirts.
	Man seduces.
	Young Woman experiments.
	Roe seduces.
"Intimate"	Young Woman fantasizes.
	Roe pleases her.
"Domestic"	Jones gloats.
	Young Woman avoids.

"The Law"	Reporters manipulate.
	Young Woman confesses.
	Prosecution manipulates.
	Defense defends.
"A Machine"	Young Woman resists.
	Priest saves.

Character Analysis:

Young Woman: kept together on the outside, concerned with her appearance, wanting more, lost, suffocating, pressured

Jones: "swell," self assured but slightly self conscious, confident in appearance, average looking, and all teeth

Telephone Girl: cute, spunky, confident, promiscuous, and flirtatious

Mother: protective, brash, conformist, caring without showing it

Roe: Confident, dapper, smooth talker, womanizer, wanderer

The director in this production was working within her area of interest. In fact, everyone on this production team was working in their area. The director drew from visual imagery and, more interesting, the sense of smell to communicate to her team. This was something that motivated the designers quite a bit. She also demonstrated the movement of the show by bringing in a workbench clamp and as she spoke she tightened it. The clamp finally closed to its tightest position exactly at the end of her presentation. This was done to illustrate her heroine's dilemma.

The scene designer immediately responded with a presentation that had industrial fans, vices, air ducts—the very constructivist version of the workbench clamp.

The costume designer followed suit in much the same way by providing Universal Product Code (UPC) armbands for all of the actors where the characters were to be scanned as they entered and exited the environment.

This production overall was successful in creating a cohesive collaboration. It was comfortable for the team as they were working in their respective areas. They were able to take a chance within a "concept" show and they all worked up to their potential.

The fact that this show did not have an assigned lighting designer ended up being a bonus. When we were searching for a new lighting design professor, we sent the director's notes on the production and the applying

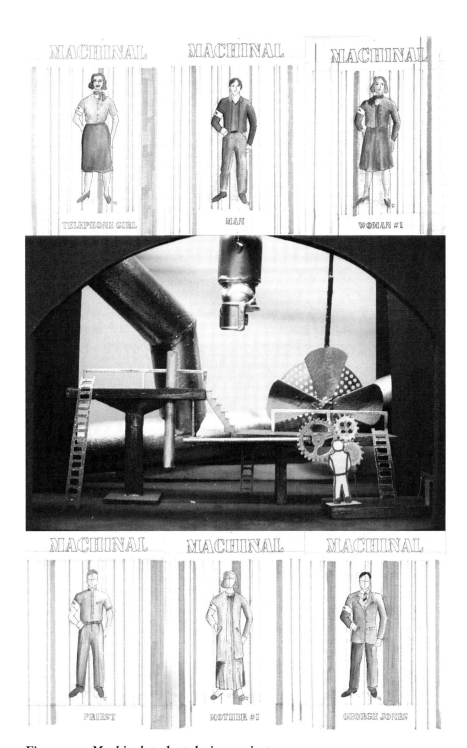

Figure 10 *Machinal* **student design projects**

Source: Scene Design by Kelly M. Leight, Costume Design by Billy Wilburn.

professors were asked to respond as the designer on this unproduced work. It was a great way to see how these prospective designers might collaborate.

Project #2: *Amadeus*

Abstract from the design team:

Amadeus

By Peter Shaffer

Concept Statement:

> *Amadeus* is a memory play, Salieri's memory. With this as our guide, the team decided to view the play through this veil of memory, and extract and heighten some details and "forget" others. To accomplish this, the set is intentionally fragmented to suggest this incomplete remembrance and reveal Salieri's focus. The lighting creates "blocks of memory," defining time and place in Salieri's mind. The costumes, inspired by Salieri's passion for sweets, support his hunger for revenge and his all-consuming desire to be "beloved by God."

In this production no one was assigned to the director's role. There were two directors working in the design areas of scenery and costumes, and a lighting designer working within his area.

Keeping in mind the strong personalities of all of those involved, each one was able to contribute to the overall concept, but their individual personalities led to three unique productions, NONE of which were cohesive.

Although they worked together swimmingly (in the spirit of collaboration), they did not affect each other collaboratively. Not having one person to funnel all the information through defeated the collaborative process. It provided for a series of dynamic ideas, all of which could be turned into unique separate design.

Project #3: *Sweet Charity*

Mise en scène from director:

Sweet Charity

Music by Cy Coleman
Lyrics by Dorothy Fields
Book by Neil Simon

Figure 11 *Amadeus* **student design projects**

Source: Scene Design by Jeanine Cull, Costume Design by Sam O'Neill, Lighting Design by Shannon Schweitzer.

Mise en scène:

I would like this production of *Sweet Charity* to be absolute and complete entertainment. Just like Las Vegas, I would like to see lots of bright colors, lights, music, dancing, and fun. I also want the play to move quickly from scene to scene, with little to no pause for set or costume changes. This is a fast pace lifestyle, the Rat Pack world. Money is flying from hand to g-string. Men are lying through their teeth for a one-night stand because what happens in Vegas, stays in Vegas. People go to dance clubs and party all night. The drinks keep coming because the bars never close. The city never sleeps. You never know who you are going to go home with, or if you are going to have to pay for it.

On the surface, and for the average tourist, Las Vegas is great and fun and entertaining. There is no need for thinking when you are playing the slot machines, watching a Vegas Revue, or ogling the various casinos. However, those who work and live in this world see a very different picture. They see the dirt and grime of a city of sin; the multiple strip joints that stink of cheap perfume and call girls mixed in with the regular crowds. The women who came with dreams of being so much more than a "dancer," but who are somehow still in the dim, filthy club where they started with wads of one dollar bills in their bras. There are cheaters and prostitutes and drug dealers and murderers, all roaming around this happy, bright, fun world.

I do not see this production focusing necessarily on the dirty side of Vegas, but I do not think one can deny that it is there in the script, especially in the men in Charity's past and her job at the Fan-Dango. I do not want to weigh down the production with really deep, heart wrenching views of the grit and grime of Vegas. Just as it is for the tourist, I want it to be presented for all its glitz and glamour. I do not want the audience to see the sleazy, cheap side of Vegas until they cannot help but run into it.

The entire play smells and looks like a cheap drink, in a fancy glass, for example, a watered down cosmopolitan, a seven and seven, or a whiskey sour.

Given Circumstances:

The play will take place in 1965 in Las Vegas, of course. It will be summer and most of the scenes will take place in the evening hours through the night. Just like Vegas, the economic situation of each character differs, and may even change from day to day. Charity and the dancers are not in the upper- or middle-class crowd because of their profession, but they probably have more money than most people in the lower class. Vittorio Vidal is definitely among the rich and famous. Oscar is in the middle, getting by very nicely.

Set:

I envision the set reflecting the world of Las Vegas as it was described in the beginning paragraph, lots of glitz and lights, big signs, and bright colors. The pieces that create each of the locations should be able to move fluidly and quickly without stopping the momentum of the play. I would like the set to stick to the "eye-candy" idea, so it would be nice to have a couple of "wows" for the audience, for example, at Coney Island or the Pompeii Club.

Costumes:

I imagine the costumes fun and colorful also, and in the period of the 1960s. While working, I see Charity and the girls in outfits that are well put together, to show off their best assets; however, they may look worn because they have been working in them for so long. I see Vidal as a sort of Rat Pack figure, because he actually is a part of that Hollywood scene; very smooth, dapper, and rich looking. Oscar is not as smooth, but I think he still has nice things. I do not picture him going out of his way to make himself noticed, so I do not think he would have on the finest clothes or necessarily the latest trend. I see Herman as a greasy, slicked back kind of guy, maybe in a cheap knock-off suit or something comparable to a skinny leather tie from the 1980s.

Choreography:

I envision the choreography to be Fosse inspired. As with the set design, I would like some of the dance numbers to "wow" the audience, like the big Vegas Revues attempted to do. Some possibilities might be in the "Rhythm of

Life Church" or "Big Spender." The dances should be fun, energetic and fast paced.

Dialogue:

Some of the girls at the Fan-Dango may choose to use a dialect, considering they have all come to Vegas from somewhere else to make it big. Vittorio will also have an accent to make him seem more exotic than he already is. Other than that, the dialogue will move at a generally fast pace, in and out of the songs.

Character Analysis:

Charity: vivacious, spunky, innocent in her thoughts and expectations, a bit naïve too, tenacious

Helene: rough around the edges, motherly, hard, a little bitter and jaded, resilient

Nickie: stuck in her world, hopeful, sexy and knows it, uses it to her advantage in every situation but worries that it might be all she has going for her

Vidal: smooth, charming, oozes money. With Charity, he is honest and open

Herman: a little slimy but he is protective of the girls, a bit of a wimp, hides behind his club, maybe has a "small man complex" because of his unimportance

Oscar: honest, needs to be taken care of, a bit of a loner, slightly nerdy

Polar Attitudes:

Charity: somehow sees the world as an opportunity, something that she will be able to take advantage of as soon as she gets that one shot

Helene: sees the world as a pessimist, she has given all she can and knows she isn't going to get any further

Nickie: sees the world as it is, but believes there is more for her; but she is also afraid of moving out into the world outside of the club where she is safe and comfortable

Vidal: sees the world as a stage, always pleasing the public, and being who they expect him to be and loving it

Herman: sees the world as basically the four walls of the Fan-Dango, his livelihood. He is content taking care of the girls and providing entertainment for his crowd

Oscar: sees the world as average and slow going until Charity comes along. She opens his eyes to her bright colored world (Coney Island). He sees the world traditionally in his values and lifestyle

Dramatic Action:

Act I

"The Park"	Charity denies.
"Fan-Dango Intro"	Charity denies.
"Big Spender"	Charity renounces men.
	Girls seduce.
"Meeting Vidal"	Vidal makes Ursula jealous.
	Charity obliges.
"Pompeii Club"	Vidal debates his situation.
	Charity listens.
"Vidal's Apartment"	Vidal forgets Ursula, then pleads w/ Ursula.
	Charity flirts, then reassures Vidal.
"Something Better"	Charity reaffirms her dreams.
	Helene mocks.
	Nickie dreams.
"The Elevator"	Charity rescues.
	Oscar wants to convince himself he is brave.

Act II

"The Elevator Cont."	Charity flirts.
	Oscar confronts.
"Rhythm of Life"	The congregation believes.
"First Kiss"	Charity evades.
	Oscar charms.
"Dream Your Dream"	Charity convinces herself that Oscar will be okay with her profession.
	Helene and Nickie dream.
"Coney Island"	Charity falls in love.
	Oscar falls in love.
"I'm Going"	Nickie mocks.
	Helene avoids.

	Charity decides to leave.
"Telegram"	Charity hopes.
"The Booth"	Charity comes clean.
	Oscar accepts.
"Surprise"	Charity celebrates.
	Oscar doubts.
"Dumped"	Charity begs.
	Oscar regrets.

This particular group had the additional stipulation of working completely in a digital design medium (in scenery, costumes, and lighting). The concept explored an updated 1960s Las Vegas in all of its neon. Working within the constrictions of the digital media pushed everyone to an uncomfortable level, including a director who had never responded to such renderings or anticipated the amount of time it takes in order to produce such renderings.

With the added digital learning curve, the director was consistently frustrated with the response time from the designers. Also in this project the scene and lighting designers had to be nearly intertwined because of the rendering program (3-D Studio Max) they were using to prepare their renderings.

The result was a little bland because of the digital constrictions. This, in turn, negatively affected the ease of the collaborative process. Because of the learning curve, the collaboration wasn't as fruitful, but you see the beginnings of a collaborative digitally designed production.

The director understood the process throughout, but did not know how to respond and didn't want to add stress to the designers. Everybody was working in a new medium, no one really knew how best to push the designer, please the director, and ultimately please themselves. All in all: pockets of success.

Student Responses

Define collaboration?

- The putting together of ideas and talents toward a common artist goal. The mutual coming together of various forms of art focused on a common artistic goal.
- Multiple people working together to enhance one common goal.
- Working together to achieve the same goal or goals.
- A group of individuals working with all their ideas in order to reach a common vision.

Figure 12 *Sweet Charity* **student design projects**

Source: Scene Design by Shelley H. Barish, Costume Design by Angela Wendelberger, Lighting Design by Shannon Schweitzer.

- Understanding and cooperation between all areas of theatre production—working together to produce the best show possible to carry out the Director's vision.
- The process of creating one unified/cohesive product through a sharing of ideas and group problem solving.

What was your overall impression of the class?

- It was a great class. I learned so much about theatre and striving toward working together with people. It is the one class every theatre practitioner should take. I wish some of the people I work with today had taken it. It taught me how to respect other people's ideas and how to get excited and find my own artistic niche within the production even if the vision was not my own.
- The class is a must for any design/directing student. It is not enough to simply have the skills such as drawing, painting, drafting, knowledge of space, and environment. Knowledge of the craft is important in any art, but in theatre, artists must also be able to communicate with each other and work toward a common goal.
- I thought this was an excellent class to have in a graduate program. It was a really super safe way to fail in the design world without ruining one of my actual designs for an actual stage.
- Very nice working with different designers and directors. Very helpful in establishing working relationships with others as well as understanding the roles of other designers and directors.
- I liked the class. It allowed the students to explore in a safe way how to work with others (some who were more demanding than others).
- It was a great experience to learn how to see the theatrical world through the eyes of the other parts of the production.
- It was a class in which true collaboration took place.
- A great exercise on the pre-production aspects of theatre.

What was the most successful element of the class?

- Learning the vocabulary needed to convey to every part of the design team what I was feeling.
- Designing outside of your area.
- I think the *experience* is the most successful. The actual designs and concepts are good too, but being able to just have the experience in a safe environment is beneficial.

- As a past director, learning to work with costumers, lighting, and set designers.
- The fact that both designers and directors were in the class.
- Learning better ways to communicate with other members of the production.

What was the least successful element of the class?

- What worked and didn't work proved the value of collaboration. If a team player was not holding up their end of the team, it pulled the whole team down and ultimately the design suffered.
- Working outside of my area.
- The inexperience of some directors added to the frustration of their designers.
- I think it was too many projects for the amount of people in the class. It needed more focus.
- Having to work as a costume designer, lighting designer, or set designer without the creative background to do so.
- I don't think there was an unsuccessful element.
- Problem solving with budget constraints.

How did you feel when working on projects out of your area of emphasis?

- It felt good. There was less pressure to get it done right and well, so I think I had more fun with it and approached it in a relaxed manner. Plus, I trusted the other people working in outside areas to not let me fail.
- I hated it. I feel that the director I was given for that project may have influenced that reaction. I happened to hate the concept, and I didn't know enough about set designing to make it work. It was hard enough working out of area.
- I found it to be a challenge, but I took pride in it since it was successful.
- I had no problem with it.
- It was frustrating not to be able to do a good job with the presentation, as I do not have artistic talents in the areas of rendering, drawing, etc.
- It was nice to be pushed into a new area and understand the process of other designers.
- Nervous and anxious, but, with help from fellow classmates and the professor, I was able to securely enter into the world of design and present my ideas.

What would you change about the class, syllabus, or deadlines?

- Again, looking back I think it was hard to focus as much on the craft-the model etc . . . when so much of the class was about ideas. Although the class gave us the opportunity to work more on the collaborative process, I think we could have used more emphasis on the craft.
- Nothing.
- It is hard when you don't have a conducive schedule for extra meetings that people want to have. We have such different schedules that it is hard to meet. There needs to be more supervision of that. Deadlines are what they are—sometimes you meet them, sometimes you don't.
- Allow those majoring in directing to be directors for most of the projects or work with a designer to do the drawings, renderings, etc.
- Having a crash course in presentation styles for each design and director, such as knowing some example way up front could have saved time through the process and more time could be spent on the actual collaborating instead of presenting. Also, time spent in class fine-detailing aspects of the design between designer/ professor could have helped and saved time spent outside of the regular class time, allowing more time for the group to collaborate (although learning how to resolve such design issues was interesting). Setting up the class similarly to a lecture/lab class would have also been better to utilize time. Having a lecture/presentation two times a week and then one lab class with a mentor or teaching assistant broken up to focus on each group would have been great with organizing the students' time better. Assistants in each field teaching would have also been better to utilize the expertise of the head teacher (in this case, Kirk Domer); instead of spending time on little design issues, he could have worked on the communication issues, and the assistant would have been able to discuss with the student separately.

Did you feel prepared to enter this class in terms of past training?

- Yes. I felt I had the tools (model making/design), but that it was about learning how to communicate with each other.
- Yes (multiple entries).
- No. A beginner course for each design aspect would (in any grad situation: especially directing) be a great pre-requisite and would be a time-saver for class. Although, having people outside of the theatre arena as part of the class was very valuable in thinking outside of the traditional "theatre box."

Were the expectations of the syllabus reasonable?

- Yes. Very reasonable. We were simply asked to do what we will need in order to be part of a production/design team.
- Yes (multiple entries).
- Yes, more than reasonable.
- Yes . . . and challenging for each student to do an aspect of a production outside of the specific area was a challenge well needed and worth participating in.

What did you gain from taking this class?

- I think the people I work with now appreciate the fact that I took this class. I learned basic communication skills. These skills can easily be translated into everyday life and dealings with people. I learned about vocabulary and just changing a phrase can be the difference between a fight and understanding. I learned that it is important to hold your own weight and that going above and beyond, even just a little bit, makes a huge difference.
- Language and an increased sense of collaboration.
- Patience and respect for the other designers and directors.
- The ability to deal with all kinds of people.
- A new appreciation for the process of working with other artistic people on the production. A better understanding of the designers' thought processes.
- The ability to communicate more effectively with many different types of designers and directors and understand the meaning of collaboration.
- I gained a better knowledge of how to communicate with each design aspect. I gained a greater appreciation for each designer's process.

Would you recommend it to others?

- I think all designers and directors should be required to take this class!
- Yes (multiple entries).
- Absolutely.
- Yes—especially directors that have never had a supportive group to work with before. It is invaluable when entering the "real world" (not that I specifically am in the real world as of yet). Also, it was a great way to recheck how I communicate with others.

What did you learn about the collaborative process?

- I leaned about how I want other people on my design team to behave and work.
- There are times when you really have to let go of what you want if it isn't doing anything for the group. You really have to be ready to work with others for collaboration to work.
- That it takes the community as a whole to achieve success.
- That it can be the most rewarding or frustrating thing one can do. That collaboration is good, but at the same time it is ok to have your own opinion.
- It is something that needs to be worked on and committed to by all parties. Collaboration is not something that just happens, it has to be actively sought after and nurtured. The skills of mediation, conciliation, and understanding must all be consciously encouraged and employed if it is to work well for all parties. The director has a huge part in making it happen.
- The ability to communicate more effectively with many different types of designers and directors and understand the meaning of collaboration.
- I learned about the give and takes that each person must deal with for the betterment of the production. I also learned that, in the end, the production must have a strong leader to drive the team. This leader must not only lead, must also create an inspiring atmosphere pushing the rest of the team to not only collaborate but to also work and create to their fullest potential.

CHAPTER 3

COLLABORATION IN LIFE

In this chapter we will explore

- graduates of the class in real situations;
- possible negative repercussions;
- possible adaptations for the course; and
- collaboration beyond theatre.

Reflections

We have seen the graduates of this studio in action in actual production meetings and the result is exactly as wished. They are, for the most part, healthy production team members. Like most emerging artists they have only to refine their process and gain confidence, but the groundwork laid by such a course has immeasurably improved the level of design as well as the entire production experience for all involved. From actors to technician, the collaborative experience has become part of an entire community vocabulary.

Students who worked in design areas other than their own in class now seem to respect the demands of their fellow designers. These graduates now have a clearer definition of their own duties as well as those of each designer. Now, the collaborative process runs more smoothly due to an experienced and shared vocabulary. This newfound knowledge impacts each element of their contribution from research through implementation.

Costume designers are now more likely to ask for a light lab to check fabrics. The request for the lab is now promptly answered because of the knowledge by the lighting designer of the importance of this event. And, finally, the reciprocal question is posed by the costume designer, "What do you need from me?"

Directors from the class now have a deeper appreciation for the work of their designers. Directors now recognize the arduous process of designers as a path different, yet completely as difficult, as their own. Directors are now more likely to find research to incite their designers. Designers in turn respond to these abstract images with more adventurous designs. Ultimately all involved are more satisfied.

On the most basic level, the knowledge and vocabulary gained in these projects has accounted for respect. On a greater level, the deep impact of experiential learning of director as designer (and vice versa) has created a bridge between these disparate disciplines. Designers and directors working together have created a mutually respectful community of artists who share an understanding of the responsibilities and demands of disciplines. The collaborative process has been completed. There is no more to be taught as educator.

The rest is up to fate...and producers.

R: Since this class, I have worked with several of the student designers and beam at what I know they learned in that course. I see their healthy manipulation of the entire collaboration as they become equals in the process. We were in a production meeting for Bock and Harnick's *She Loves Me*, in which one of the past students was designing both costumes and scenery. His constant desire to keep his fellow collaborators and assistants in the loop was heart-warming. One day, while sharing his research, he pulled out an image and handed it to the lighting designer saying, "I heard what you were talking about at the last meeting and I found this image. I thought you might like it."

K: I found the same to be true in regard to the excitement that the students share when successful collaboration takes place. They are so happy when their voices are heard. Those students are also teaching others. They pass this information along to up-and-coming designers about "the way things should be."

Approaching New Students

The positive feelings created by this course also have a flip side. When collaboration is not a priority in the design process, students are the first to point fingers. Their understanding of the "best way" to collaborate makes for a team that demands the understanding of the theories in this book.

Students experienced in the process expect all involved to collaborate willingly and effectively. Some theatrical artists simply do not work that way. These students may be unprepared for such a realization. While their naiveté may be problematic now, through experience they will be able to work effectively with any artist by adapting their definitions of collaboration. By leading through example and making sure they always enter the process with their collaborative filter engaged, students may affect their production team. In working with these students perhaps dictatorial directors and facilitating designers may be transformed.

Collaboration is about adaptation and leadership within your best artistic sense.

Evolution of the Class

As always, collaboration thrives through process and evolution. This class is no exception. Our most recent incarnations of the class included actors. These actors performed scenes in the style of the productions developed in class collaboration. This new element was an added communication hurdle for the directors and designers. The aim was to ensure that everyone involved were part of the same production and succeeded as actors gained a deeper appreciation for the design aspect of the production and vice versa.

The actors became sensitive to correctly inhabiting a stylized world. The designers had actual inhabitants for which to design. And the directors spent time honing their communication skills for both actor and designer.

It is hoped that the benefits will multiply exponentially when introducing a new member into the equation.

Your class may develop in similar ways: perhaps including your theatre administration students to create publicity and marketing strategies for an imagined production. Perhaps including your dramaturgy and history students to assist in research for everyone involved in the process. Any person who is touched in the production process can be involved . . . including audience members.

No matter who you recruit for this class, it is imperative to have your students operate, at times, outside their area of expertise. Their learned understanding and compassion for their new positions is invaluable.

Collaboration beyond Theatre

Our collection of ideas extends well beyond the wings of the stage or the walls of the classroom. Our hypersensitivity to the process while writing this book has opened our eyes to the numerous daily collaborations one may encounter. Collaboration rears its head in the most mundane of situations:

> *What should we do tonight?*
> *Where should we eat?*
> *Do you want me to drive?*

Each query poses a collaborative possibility. We have found that you may answer these sorts of mini-collaborations using the ideas within this book. Though your friends may balk at the shared in-depth process of answering these harmless questions, you WILL discover what WE want to do tonight, where WE should eat, and WHO should drive.

From meetings to group e-mails, the necessity for this refinement has been proven. In real life situations devoid of theatre, collaboration theory has affected us in personal, familial, and professional relationships.

Stressful personal situations have been calmed by learning to adapt our vocabulary.

Visits home during holidays are all about collaboration.

Difficult professional situations have been salved through working as a team.

Collaboration in life.

Final Thoughts

This final part of this book aimed to synthesize the theory and practice into the classroom in hopes of affecting willing collaborators who have the wherewithal to explore the process. Whether through the theory covered in part I, the practical examples of *The Life*, or within the classroom—our main goal was to offer suggestions to manage, maintain, and mature your inherent collaborative skills.

The tools offered in this book aimed to smooth any tensions or road-blocks hindering a fruitful creative process. Although we may have failed at certain moments in our own projects, you have hopefully have gained insight to create a truly collaborative production team: a team willing to sketch and scream over a drawing on a cocktail napkin.

BIBLIOGRAPHY

Works Referenced

Anderson, Robert. *Tea and Sympathy*. New York: Random House, Plays, 1953.

Auburn, David. *Proof: A Play*. New York: Faber and Faber, 2001.

Ball, David. *Backwards and Forwards: A Technical Manual for Reading Plays*. Carbondale, IL: Southern Illinois University Press, 1983.

Bernstein, Leonard, Arthur Laurents, and Stephen Sondheim. *West Side Story: A Musical*. New York: Random House, 1958.

Bock, Jerry, Jerome Weidman, George Abbott, and Sheldon Harnick. *Fiorello: A New Musical*. New York: Random House, 1960.

Bock, Jerry, Joe Masteroff, and Sheldon Harnick. *She Loves Me: A Musical Comedy*. New York: Dodd, Mead & Company, 1963.

Bock, Jerry, Joseph Stein, and Sheldon Harnick. *Fiddler on the Roof*. New York: Crown, 1964.

Churchill, Caryl. *Cloud Nine: A Play*. London: Pluto Press, 1980.

Coleman, Cy. *The Life*. New York: Tams-Witmark Music Library, 1996.

Coleman, Cy, Betty Comden, Adolph Green. *On the Twentieth Century*. New York: Samuel French, 1980.

Coleman, Cy, Larry Gelbart, and David Zippel. *City of Angels*. New York: Applause Theatre Book Publishers, 1990.

Egri, Lajos. *The Art of Dramatic Writing: Its Basis in the Creative Interpretation of Human Motives*. New York: Simon & Schuster, 1946.

Gershwin, George, Ira Gershwin, and DuBose Heyward. *Porgy and Bess*. New York: Gershwin, 1935.

Hellman, Lillian. *The Children's Hour*. New York: New American Library, 1962.

Howard, Pamela. *What Is Scenography?* London and New York: Routledge, 2002.

Link, Peter. *King of Hearts* [Sound Recording], Music by Peter Link, Lyrics by Jacob Brackman. Englewood Cliffs, NJ: Take Home Tunes & Original Cast Records, 1992.

Loesser, Frank, Jo Swerling, Abe Burrows, and Damon Runyon. *Guys & Dolls: A Musical Fable of Broadway*. New York: Frank Music, 1953.

Marks, Walter, Ernest Kinoy, and Joseph Mitchell. *Bajour: A Full Length Musical*. Chicago: Dramatic, 1976.

Molière, Jean-Baptiste Poquelin. *Tartuffe: Comedy in Five Acts*. Trans. into English Verse by Richard Wilbur. New York: Harcourt, Brace, & World, 1963.

O'Brien, Richard. *The Rocky Horror Show: Musical*. New York and London: Samuel French, 1983.

Ravel, Maurice and Sidonie Gabrielle Colette. Trans. into English by Katharine Wolff. *L'Enfant et les Sortilèges*. Paris: Durand, 1932.

Rebeck, Theresa. *Bad Dates*. New York and London: S. French, 2004.

Rodgers, Richard and Oscar Hammerstein. *Oklahoma!* London and New York: Williamson Music, 1943.

———. *Carousel*. London: Williamson Music, 1945.

Shaffer, Peter. *Amadeus: A Play*. London: Deutsch, 1980.

Shakespeare, William. *Twelfth Night; or, What You Will*. Cambridge and New York: Cambridge University Press, 1949.

———. *Romeo and Juliet*. Ed. Jill L. Levenson. Oxford and New York: Oxford University Press, 2000.

Simon, Neil. *Barefoot in the Park: A New Comedy*. New York: Random House, 1964.

Simon, Neil, Cy Coleman, and Dorothy Fields. *Sweet Charity*. New York: Random House, 1966.

Smith, Evan. *The Uneasy Chair: A Cautionary Tale in Three Volumes*. New York: Dramatists Play Service, 1999.

Sondheim, Stephen. *Sweeney Todd, the Demon Barber of Fleet Street*. New York: Applause Theatre Book Publishers, 1991.

Strauss, Johann. *The Queen's Lace Handkerchief*. New York: Arthur W. Tams, 1898.

Treadwell, Sophie. *Machinal*. London: N. Hern, 1993.

Ward, Robert, James Stuart, and Richard Edward Connell. *A Friend of Napoleon: An Operetta in Two Acts based on the Short Story by Richard Connell*. Boston: Vireo Press, 2005.

Williams, Tennessee. *A Streetcar Named Desire: A Play*. New York: New Directions, 1947.

———. *Summer and Smoke: A Play*. New York: New Directions, 1948.

Wilson, Lanford. *The Hot L Baltimore*. New York: Hill and Wang, 1973.

Supplemental Reading

Ball, William. *A Sense of Direction: Some Observations on the Art of Directing*. New York: Quite Specific Media Group, 1984.

Clurman, Harold. *On Directing*. New York: Simon & Schuster, 1972.

Dolman Jr., John, Richard K. Knaub, and John Jeffery Auer. *The Art of Play Production*. New York: Harper & Row, 1973.

Essig, Linda. *Lighting and the Design Idea*. New York: Harcourt Brace, 1997.

Gillette, J. Michael. *Theatrical Design and Production: An Introduction to Scene Design and Construction, Lighting, Sound, Costume, and Makeup*. Boston: McGraw Hill, 2000.

Glenn, Stanley L. *A Director Prepares*. Encino, CA: Dickenson, 1961.

Hayes, David. *Light on the Subject: Stage Lighting for Directors and Actors—And the Rest of Us*. New York: Limelight Editions, 1989.

Hodge, Francis. *Play Directing: Analysis Communication and Style.* Englewood Cliffs, NJ: Prentice-Hall, 1971.

Jones, Robert Edmond. *The Dramatic Imagination: Reflections and Speculations on the Art of the Theatre.* New York: Theatre Arts Books, 1969.

Miller, Scott. *From Assassins to West Side Story: The Director's Guide to Musical, Theatre.* Portsmouth, NH: Heinemann, 1996.

Peithamn, Stephan and Neil Offen. *Stage Directions: Guide to Directing.* Portsmouth, NH: Heinemann, 1999.

INDEX

Note: References to illustrations are in **bold**.